Tynchtyk Mukanov

Investigation of sediment deposition in reservoirs of hydroelectric installations

Tynchtyk Mukanov

Investigation of sediment deposition in reservoirs of hydroelectric installations

ScienciaScripts

Imprint
Any brand names and product names mentioned in this book are subject to trademark, brand or patent protection and are trademarks or registered trademarks of their respective holders. The use of brand names, product names, common names, trade names, product descriptions etc. even without a particular marking in this work is in no way to be construed to mean that such names may be regarded as unrestricted in respect of trademark and brand protection legislation and could thus be used by anyone.

Cover image: www.ingimage.com

This book is a translation from the original published under ISBN 978-3-659-86525-1.

Publisher:
Sciencia Scripts
is a trademark of
Dodo Books Indian Ocean Ltd. and OmniScriptum S.R.L publishing group

120 High Road, East Finchley, London, N2 9ED, United Kingdom
Str. Armeneasca 28/1, office 1, Chisinau MD-2012, Republic of Moldova, Europe
Managing Directors: Ieva Konstantinova, Victoria Ursu
info@omniscriptum.com

Printed at: see last page
ISBN: 978-620-8-53763-0

Table of Contents

General characterization of the work

Relevance: The study is devoted to one of the most topical and insufficiently covered up to now issues - accurate determination and, especially, forecasting of losses of channel reservoir capacity and the issue of sediment disposal in irrigation and power reservoirs.

The Central Asian region is one of the economic areas where the basis of economy is irrigated agriculture and energy, which is based on the joint use of water resources of the Amu Darya and Syr Darya river basins. Favorable natural conditions as mountainous areas of our country as Kyrgyzstan, Tajikistan and rich water and land resources of Uzbekistan and Turkmenistan have created great opportunities for the development of energy and irrigated agriculture, the efficiency of which is directly related to water availability. In these low-water areas, especially in the republics of Uzbekistan and Turkmenistan, where relatively small and unstable flow of local rivers does not provide any reliable supply of water to consumers, only construction of reservoirs allows meeting the needs of irrigation and water supply. Therefore, construction of reservoirs in this region is a vital necessity, despite the negative consequences of this (flooding of large areas, water losses for filtration, evaporation and siltation, erosion of the channel in the lower reach by clarified water, lack of channel colmatization due to lack of suspended sediments, impossibility to use fertile impurities dissolved in water in fields, etc.).

In the Aral Sea basin, more than 60 reservoirs with a usable water volume of more than 10 million meters^3 each have been constructed. The full total volume of the reservoirs is 64.5km^3, of which the usable volume is 46.5km$^{(3)}$, including 20.2km(3) in the Amudarya River basin and 26.3km^3 in the Syrdarya River basin.

Purpose of the research: Improvement of technological operation of structures and reservoirs and consideration of methodology for determining and forecasting the volume of siltation of reservoirs: Orto-Tokoi, Kambarata HPP-2 of daily regulation by analyzing the available field measurements in the site which is located between Kambarata HPP1 and HPP-2 at the gauging station of Uchterek village. Uchterek.

Objectives of the study:

- ensuring a given degree of water purification from sediment;
- the accident-free passage of floods, driftwood and fins;
- Analysis of the sedimentation process and the effect of additional load on the pump house;
- dam maintenance and development of measures for the operation of hydraulic structures and reservoirs.

Object of the study**:** The object of the study is water reservoirs of Kyrgyzstan, in particular Ortotokoi,

Uchkurgan, Toktogul and Kambarata HPPs.

Research methods: The solution of the set tasks was carried out on the basis of collection, system analysis and generalization of field studies on the Naryn River in the sites near Uchterek village, upstream of the mouth of the Kekemeren River and the Kekemeren River in the mouth site.

Scientific and practical significance of the research results.

The work is based on extensive factual data on the study of siltation of large irrigation channel reservoirs carried out by SANIIRI, Bathymetric Center under the MAWR.

Kambarata HPP-2 is not only a source of electric power generation but also an experimental HPP. Kambarata HPP-2 was chosen by the designers specially, at the place of tectonic fault "Yuzhny" to study tectonic impact on water conduits №1,2,3, construction-operational spillway and ground dam built by directed explosion.

The proposed method of calculation of siltation of the reservoir of Kambarata HPP-2 will allow to forecast the volume of the bowl and terms of operation of the daily regulation reservoir.

The results of the research are categorically necessary for operation not only of the reservoir, but also of hydro unit No.1, which is already ready for operation and two subsequent hydro units No.2, No.3 of Kambarata HPP-2.

Brief results of the research: literature sources on the experience of operation of reservoirs of hydroelectric complexes were analyzed. The problems of reservoir siltation, i.e. the process of sedimentation of the reservoir bowl and reformation of its banks, i.e. change of the original shape of the coastal slopes due to destruction of the overwater part of the slopes by waves and formation of coastal shoals, were considered. The experience of operation of reservoirs of hydrosystems has been studied and generalized analysis has been carried out.

Scientific novelty of the work: according to the results of the conducted researches the development of recommendations on reservoir bed cleaning and the main measures on increasing the efficiency of hydrosystems operation and increasing their production output will be carried out.

Publications on the topic of Master's thesis: 1 article was prepared and published.

Key words: analysis of river water intake condition; technical operation; rational use; river water intake structures; field observations and researches; sediment accumulation and sediment flushing; hydraulic regime of water intake structures operation; flow regulation; river discharge; turbidity; fractional composition; type of layouts; manipulation of gates on multi-span structures; technological functioning of water intake structures, technological limitations of river water intake and water regulation.

Chapter 1: Hydrological regime of rivers and river processes

1.1 Rivers and their water regimes

Rivers are fed by snowmelt (spring), rainfall (summer and fall), and groundwater (summer and winter). The amount of water in a river fluctuates from year to year and within a year. The water level in the river, which is measured relative to some notional horizontal reference surface, also varies accordingly. Water regime of the river is a change in time of water level in the river, caused by the processes of accumulation and consumption of moisture in the river basin. Characteristic states of the water regime are called its phases. The main phases of the river water regime are: flood, flood, low water period. The flood is repeated annually due to seasonal climate change and is characterized by a high and prolonged rise in water level caused by snowmelt in the river basin. During floods, water leaves the river channel and floods its floodplain. On the rivers of Kyrgyzstan, floods occur in the spring months of March, April and May and can last from two to three weeks to one to two months. Floods can occur repeatedly and irregularly in different seasons of the year and are characterized by an intense short-term increase in water level. A flood that is outstanding in magnitude and rare in recurrence, which may cause casualties and destruction, is called catastrophic. A flood with the same consequences is sometimes called such. Flood is the inundation of an area with water, which is a natural disaster. Flooding can be observed during floods or floods, when water rises due to wind surge of water at the river mouth, due to the breakdown of hydraulic structures.

The low water period is annually repeated in the same periods of the year and is characterized by a prolonged low water level resulting from a decrease in the river's supply. A distinction is made between summer and winter low water. For the beginning of the summer low water is taken the end of the flood, for the end - the onset of water rise after the fall rains. Usually the low-water period falls on July-August, when the rivers are fed mainly by groundwater. By constructing reservoirs on rivers, the water regime can be effectively regulated throughout the year. As a result, it takes place at relatively equal water levels.

The flow of water in rivers is caused by the gradient of the water surface from source to mouth: the greater the gradient, the higher the flow velocity. This velocity also depends on the average depth, sinuosity and roughness of the channel. It is different on the spits and rapids: in low water it is higher on the rapids, in high water - on the spits. Flow velocity decreases from the source to the mouth of the river. The flow velocity of a river is the average flow velocity of individual jets of the river stream. The velocities of jets are not the same at different points in the cross section of the stream, the area of which is called the live section of the river channel. The jet velocity is maximum over the deepest place near the water surface at a point about 0.2 depth from the surface. From this point, the jet velocity decreases toward the river bottom and its banks. At the water surface, the velocity of the jets is maximum above the deepest point. The line connecting the points on the water surface with the highest jet velocities and the greatest depth of the channel is called the river stem. At the bends of the channel simultaneously with the flow along it is observed and transverse flow: at the surface - towards the concave bank, at the bottom - towards the convex. So on curvilinear sections of the river, the water flow seems to "screw" along the channel downstream. This screw-shaped water flow "carries" the river stem closer to the concave bank. Thus, on rectilinear sections of the river the flow velocity is highest in the middle of the channel, on curvilinear sections - closer to the concave bank. When the water level changes, two additional transverse currents appear between the middle of the river and its banks. When the water rises, these currents on the water surface are directed from the middle to the banks, on the bottom - from the banks to the middle, and in the middle of the river - from bottom to top. When the water is receding, it is the other way around. All rivers are characterized by sediment deposition. The most characteristic type of sediment deposition is the spit, which forms at a convex bank in the form of a wedge running at an angle downstream. The spit gradually goes under water, extending far into the river channel. Sediment deposits in the channel can be either underwater or overwater. Rolls. This is a stable sediment deposit in the form of a berm crossing the riverbed (often called a ford). It is the main factor complicating navigation on rivers. Most often rolls are located in places where the water flow transitions from one bend to another. In high water, flow directions are parallel to the banks of the rolls, in low water - depending on their features. For example, the formation of a swamp current depends on the shape of the basement: for example, on an overkat with a flat basement, the well-defined river stem coincides with the trough axis and there are no swamp currents. Solid material is transported by rivers in three ways: dissolved, suspended and dragged along the bottom. The first method plays almost no role in the formation of placers, the second one plays a very subordinate role (in the case of transportation of gold), and the third one is the main one. The main mass of solid material is carried by the river during floods. At this time the amount of water and the speed of its flow are so great that the river sets in motion a considerable layer of sediment covering its bottom. This layer of moving sediment is usually several decimeters thick and sometimes 1-2 m thick. It consists of pebbles immersed in a liquid mush of water, sand and muddy particles, i.e. a kind of mud flow. In its lower parts it has a thicker consistency, which gradually becomes more liquid in the upper parts of the layer. In the same way, the speed of movement changes: it is minimal (up to zero) in the lower parts of the layer and maximum at the top. The larger the flood, i.e. the greater the water mass and the greater the current velocity, the greater the power of the moving bedload layer. The movement of pebbles and larger particles is not so much by rolling as by sliding one layer over another. It is quite clear that in this movement valuable metals, as well as increasingly heavier minerals, inevitably find their way to the lowest parts of the moving layer and are concentrated in the slowest moving part of the layer, which is only a few centimetres thick. Thus, already in the very process of solid material transportation, two layers are separated, with and without heavy minerals, which are like prototypes of sands and peats. As soon as the high water subsides, i.e. its mass and flow velocity decrease, the movement of the bottom sediments is immediately weakened. The lowest part of them, the proto-sands, so to speak, becomes immobile, fixed. As the water recedes, the upper parts of the moving layer become fixed and its power decreases. During low water, the movement of the bottom sediment almost ceases and takes place mainly by rolling over the bottom of individual pebbles. In the next flood, the sediment is again in motion. If the flood is of the same or greater magnitude than the previous one, the layer containing the valuable metal will move. If the flood is of lesser magnitude, this layer does not move. Even in this case, however, the lowest part of the moving bedload layer may contain valuable metal, provided that it is brought in from upstream. When the high water recedes, this new metal-bearing layer will also build up on top of the previous one. If a given stretch of stream is in the stage of sediment accumulation, i.e. the amount of material brought in is greater than that taken out and the total thickness of the sediment is gradually increasing, the metal-bearing layer will build up from year to year by fixing the lowest part of the moving bottom sediment. Accumulating over the course of centuries, this accumulation leads to the formation of a more or less thick layer of metal-bearing sands. The accumulation of sands lasts until the influx of valuable metal from upstream parts of the river stops. After this, peat, i.e. the same sediment but without the valuable metal content, accumulates in the same way. If all floods were of exactly the same magnitude, then each year an equal, very small part of the bedload layer would be fixed, corresponding to the general rate of accumulation of sediment in a given section of the river. But there are usually alternating high and low floods, and even whole periods of both. Exceptionally high water can draw a long-established bedload layer into motion, over-wash it and redistribute valuable metal within it. If such a high flood does not occur again, or if it occurs after a long interval when the total thickness of the sediment has increased considerably, this layer will remain permanently immobile. The layer from the next exceptionally high flood will be deposited on top of it in the same way. Thus the river load that composes the alluvial placers is not the sediment of average or even high waters, but only of the exceptionally high floods experienced by the river during the period of sediment accumulation. This is the principle of the gradual accumulation of river-bottom contributions.

During floods, when flow saturation and velocities reach their highest values, channel scouring is observed in some places, which is particularly intense on concave banks at river bends, where the influence of the lateral circulation is felt. The eroded material is deposited on opposite shoals and is partially carried away to the rolls, the bottom of which is thus considerably increased during floods.

However, scouring processes do not stop even after the passage of floods, but the opposite phenomenon is observed: local gradients and significant gradients are formed on the tumbleweeds when the water recedes, causing scouring and sediment transport to the underlying channels. This process, in the practice of river studies, is called the process of self-cleaning of the river crossing.

The greater the gradient of rivers, the more intensive the processes of scour and deposition, and the greater the relative amounts of sediment transported by the rivers. In particular, the most intensive scouring of the bed and banks of the channel occurs in the upper reaches of rivers, where flow velocities reach their highest values and floods rise most intensively.

At present it can be considered as generally recognized that the processes of transport of solid particles in suspension by the flow are a direct consequence of turbulence of the moving fluid. In a laminar flow there is no reason for the occurrence of suspension - all solid particles, caught in such a flow, will inevitably sink to the bottom, and stop in their movement or move along the bottom.

In turbulent flow, in addition to the main motion in the longitudinal direction, there are additional transverse movements of liquid masses, the presence of which is the main reason for the transfer of solid particles from the lower to the upper layers and transport them in suspension over considerable distances. Since solid particles have a higher specific gravity than water and have a steady tendency to settle to the bottom, to keep them in suspension, it is necessary to have constant impulses in the bottom layers of the flow, causing the rise of solid particles from the bottom to the upper layers of water. As is known, the solid wall of the flow possessing a certain degree of roughness (especially in the river channel characterized by the presence of ridges) is a source of turbulence; in the bottom layers of the flow there is vortex formation, intensified by the presence of ridges, as a result of which there are force effects on solid particles, about which were discussed above.

Modern experimental studies show that behind each roughness bump or ridge a breakaway region is formed where the pressure is less than hydrostatic pressure. On the contrary, there is always an increased pressure on the frontal side of the particles protruding from the bottom. Non-symmetric streamline of particles lying on the bottom leads to the appearance of a significant lift force, the value of which can be determined on the basis of the well-known theorem of N. E. Zhukovsky.

Under the influence of frontal and lifting forces the stability of solid particles is broken, they begin to move along the bottom, and at certain velocities they are involved in the flow column.

The interaction between solids and flow under given hydraulic conditions always leads to some when the number of years of observations is clearly insufficient (e.g. if the observation period does not cover high or low water years). However, in case of insufficient observation data, it is preferable to use the method of analogies to construct daily flow availability curves, which allows to lengthen short series and obtain sufficiently reliable data. Empirical formulas do not fully take into account the factors affecting the patterns under study and therefore cannot be considered reliable. As for the method of analogies, it has been repeatedly used in the practice of hydrological calculations and can be considered satisfactory enough.

Sediments are carried by river flows, both in suspension and by dragging along the bottom. In mountainous and foothill areas, where slopes often exceed critical gradients, rivers are able to carry coarse pebbles, cobbles and gravel mixed with sand of various fractions. At the same time, they carry in suspension significant masses of finer sand, silt and clay. In flat areas, where gradients decrease sharply, pebbles and gravel are much rarer and the bulk of the load is made up of sandy-silt particles of various sizes. In the lower reaches of rivers, where the current slows down and the transport capacity of the river flow is significantly reduced, sediment deposition occurs, leading to the formation of a delta at the point where the river meets a lake or sea.

Turbidity is the amount of suspended sediment per unit volume of water. Turbidity is measured in g/l or kg/m^3.

The solid flow rate is the amount of sediment carried through the live section of a stream in one second. Solid flow is expressed in kg/s.

The solid discharge of a river in a given cross-section is the amount of sediment transported by the stream over a given period of time (month, year and so on). The annual sediment load of large rivers is measured in millions of tons. In flatland rivers, transported sediment is a minor part, about 1-5% of the sediment load.

Water turbidity is taken into account when selecting the location and elevation of water intake from the river. The highest turbidity, as a rule, in plain rivers takes place in spring during floods. In small rivers, the highest water levels, highest discharge and highest turbidity coincide in time. In large rivers, the maximum turbidity is slightly ahead of the maximum discharge. Thus, in large flatland rivers, the maximum suspended sediment load occurs during the rise of the spring flood.

Solid flow is determined to assess the conditions of reservoir siltation and to assign a dead volume intended for siltation during the design period of reservoir operation. In some cases, it is necessary to ensure the transit of sediment from the reservoir to the downstream side of the reservoir when designing hydroelectric schemes.

Seasonal, annual and multi-year regulation is carried out depending on the reservoir's usable volume capacity and average annual runoff values.

Annual regulation serves for redistribution of flow during the year. Under annual regulation, water is accumulated in the reservoir during floods, since usually during this period the consumption discharge is less than the inflow discharge. The accumulated water volumes are used in addition to the domestic flow during low-water periods of the year.

Multiyear regulation serves for redistribution of domestic flow over several years, with water reserves in the reservoir being accumulated in high-water years and discharged in low-water years. Thus, under multi-year regulation, the full cycle of reservoir level fluctuations has a duration of several years.

Reservoirs capable of multi-year regulation can usually meet the requirements of annual regulation because less usable reservoir volume is required to implement annual regulation.

Reservoirs of annual regulation usually carry out daily and weekly regulation. Relative values of useful volumes of reservoirs, expressed in fractions of average annual runoff, necessary for annual and perennial regulation are approximately 0.1 - 0.3 for annual regulation, and more than 0.3 for perennial regulation. The volume required for daily regulation is usually about 0.1 - 0.3 of the daily runoff.

Multi-year regulation is an additional release of water from the reservoir in low-water years.

Seasonal regulation is caused primarily by irrigation needs, weekly and daily by the schedule of power generation.

Daily and weekly regulation downstream of the dam creates a zone of sharp water level fluctuations during the day and throughout the week. Under diurnal regulation, within 5-10 km of the dam, the lowest water levels are observed in the early morning hours and the highest in the evening hours. With distance from the dam, daily fluctuations in water level decrease, and the timing of rise and fall of water is shifted by the amount of time it takes for discharged water to reach the dam.

Under weekly regulation, minimum water levels occur on Saturdays and Sundays. At distances of 30-40 kilometers the dates of minimum levels can be observed on Sundays and Mondays. During low water levels fluctuations can be felt at a distance of up to several tens of kilometers, and at much smaller distances during high water and floods.

Water level fluctuations in the reservoir itself are determined by its water balance, i.e. the ratio of inflow to outflow, the degree of flow regulation by the dam, and the effect of wind on the mass of water in the reservoir. Water level changes due to water balance and seasonal regulation are up to several meters per year. Usually the water level rises in spring when the reservoir fills with melt water. During the summer the level decreases.

Weekly regulation causes more noticeable level change in the lower part of the reservoir than daily regulation. Prolonged and strong wind can cause the water mirror in the reservoir to be skewed: water surge is observed in one part of the reservoir and runoff in the opposite part. This phenomenon is more pronounced in shallow reservoirs. The highest level rise occurs at the beginning of the surge. At this time water level fluctuations can reach up to 1m.

1.2 Nanos and channel regime

The flow of a river is accompanied by the erosion of some parts of the channel and the deposition of sediment in other parts. The movement of sediment (products of channel erosion and soil erosion in catchments) depends on the discharge and velocity of the channel flow. A distinction is made between suspended sediment, which is transported throughout the entire thickness of the stream, and bed load, which is predominantly larger particles transported in the bottom layer of the stream and constitutes the main material for the formation of channel formations. On flat rivers, the bottom relief is an alternation between ridges of bottom sediment crossing the channel (rerolls) and deep areas - splays. The distance between the source and the mouth is the length of the river. The kilometers are counted from the mouth. Winding influences the counting of kilometers on the map. The degree of winding of the river is characterized by the coefficient of winding, which is the ratio of the length of the considered section of the river along the fairway to the length of the straight line connecting the ends of this section. This coefficient is greater in small rivers than in large rivers.

The underwater topography of reservoirs is shaped by several factors: the morphometry and hydrodynamics of the reservoir, the morphology of the submerged topography, and the volume and composition of sediment from runoff and bank recycling.

Reservoirs accumulate most of the material arriving with surface runoff as a result of bank collapse. No more than 4-10% of all incoming sediment is discharged downstream and therefore at least 90-95% is used to form bottom sediments

The source of material input is categorized into 5 groups:

- surface runoff;
- erosion of banks and bottoms;
- production of phytoplankton and higher aquatic vegetation;
- physical and chemical processes in water bodies;
- ash processes.

Whereas in natural inland reservoirs the load balance is dominated by material from surface runoff, in most plain reservoirs in the first years of their existence the bottom sediments are formed mainly from the products of bank and bed failure; over time their quantity decreases and the load from surface runoff begins to predominate.

Peat bogs are a specific source of bottom sediments; the peat suspension formed by them is involved in the formation of bottom sediments. In the first 25 years of the Rybinsk reservoir's existence, this amounted to about 5% of the total sediment load. The decay products of phytoplankton and higher plants are also a significant source of sediment replenishment.

In arid zone reservoirs located in sandy deserts, gross transport of sandy and dusty sediments can be an important source of sediments.

Temperate and tropical reservoirs generally do not have such a source.

The amount of substance entering desert reservoirs is significant, and that is why they silt up very quickly. Thus, in the USA, some reservoirs located in the desert zone were filled with sediment for 10-15 years; one of them lost about 50% of its capacity in 7 years. To replace this reservoir, a new one was built, but it lost 95% of its capacity.

The sediment entering the reservoir is distributed over the reservoir bed and forms a ground sediment complex. The composition and distribution of bottom sediments in reservoirs can be very diverse and depends on the size of the drawdown and the degree of flow of the reservoir, the regime of currents and waves, and the quantity and coarseness of the sediment. These factors are strongly influenced by the geographical location of the reservoir, its size and morphology. In each particular reservoir, the nature and quantity of the soil-forming material and the hydrodynamic activity of the water masses are determining factors.

In channel reservoirs the predominant form of hydrodynamic activity is currents, in valley and lake type reservoirs - waves. During the study of reservoirs bottom soils were divided into three main groups:

а) primary flooded soils that have not significantly changed their properties;

б) transformed flooded soils that have largely lost their former properties;

в) secondary soils (gravels, sands, silts, the formation of which is preceded by pre-sorting of their particles, and deposits from macroliths).

Pebbles, gravelly and coarse-grained sands are developed mainly in mountainous and foothill reservoirs, mainly in the deltas of rivers flowing into the reservoir. Sands by formation can be divided into channel, coastal and open space sands, silty sands are formed at the boundary of sand distribution, where changes in hydrodynamic activity are observed. Sandy and gray silts are divided by their origin into alluvial and local silts; the former are formed from alluvial sediments brought into the reservoir, the place of their accumulation being the zones of backwater wedging, while local silts are formed mainly from the products of erosion of the shore and reservoir bed. Remains of peat, submerged terrestrial vegetation, products of wind transport, and remains of aquatic fauna and flora are involved in the formation of all sediments.

Depending on the hydrodynamic activity (currents, waves), the depth of the sands at these reservoirs varies.

The process of underwater relief reorganization starts simultaneously with reservoir filling. Scouring and re-deepening of the bottom by currents is more often observed in channel reservoirs. Scouring as a result of wave activity usually occurs in water areas with depths not more than 8-10 m. Bottom leveling, both in shallow and deep water zones, occurs due to sediment deposition in relief depressions.

Suspended material from the agitation zone enters the open part of the water body, where it is deposited on the bottom, forming a layer of sand-silt or silt sediments. Their greatest thickness is observed in closed depressions (flooded lakes, oxbows and interdrift depressions). The thickness of sediments decreases with distance from the shore and towards the dam.

The process of reservoir bed formation, as well as the process of bank formation, is divided into two periods:

- submarine landform formation, accompanied by more or less sediment input, and submarine landform stabilization, which occurs when sediment input to the reservoir is moderate;

In the second stage, shallow shores also intensively collapse and the main elements of the bottom relief are formed; in shallow water areas, wave leveling of the bottom begins, and in the deep-water zone, primary relief leveling begins. The duration of this stage may exceed 2-4 decades.

The second period is characterized by a marked decrease in the input of material from abrasion and an increasing role for solid runoff from the catchment and the products of aquatic life. During the first stage of this period, the processes of sediment transport along the shore and the detachment and filling of bays play a leading role in the formation of the reservoir bed; there is also a slow overgrowth of shallow waters and the formation of a complex of bottom sediments is completed. During the second stage, the relief of the reservoir bed is formed by the accumulation of accumulative material further and further from the shore, which contributes to the gradual detachment of some areas of the reservoir and1 the leveling of the bottom relief. The duration of this stage may reach several hundred years.

Significant changes in formation of bottom sediments and, consequently, bed relief occur in case of reservoirs arrangement in a cascade. In case of continuous arrangement of reservoirs, the amount of solid runoff decreases noticeably towards the river mouth. The main receiver of solid river runoff is the upper reservoir, where it is accumulated. All downstream reservoirs receive suspended sediment only during the transit discharge of spring floods through the cascade and from tributaries. In this case, the inflow of solid runoff into the river delta is drastically reduced.

By accumulating the vast majority of sediment load, reservoirs, and even more so reservoir cascades, drastically reduce the amount of sediment entering deltas or estuaries.

The creation of reservoirs changes the regime of sediment movement. The nature and extent of these changes depend on many factors: the size of the reservoir, its outline in plan, the size of its drawdown, the quantity and coarseness of the sediment carried by the river, the extent of bank recycling and the flow capacity of the reservoir, i.e. the exchangeability of the water in the reservoir, expressed as the ratio of the average annual flow to the total volume of the reservoir

Due to the dramatic reduction in flow velocity, concentrated deposition of sediment, usually 0.25mm and above, occurs in the tailrace of the reservoir, similar to that observed in river mouths. The finer sediment is deposited throughout the reservoirs and is partly carried downstream. When the reservoirs are drawn down, the point where the backwater curve wedges out moves in the direction of the dam, so that the previously deposited sediment is eroded and moved in the same direction, and when the reservoir is filled the point of deposition moves upriver again. The periodic relocation of the site of concentrated sediment deposition leads to the filling of the Dead Volume of the reservoir. The reworking of the banks and alongshore sediment transport also contributes to this.

In some cases, landslides and bank collapse have a significant impact on the sedimentation and siltation of mountain reservoirs. Creation of deep reservoirs with large (up to 50-80 m) and comparatively fast level drawdown leads to disturbance of stability of their steep height banks. These violations consist of the following:

- weighing of rocks during filling and their weighting (due to high water saturation) during discharging;
- weakening of rock strength and reduction of its shear resistance as a result of variable water saturation and drying, weathering, leaching, etc;
- development of hydrodynamics of water pressure on rocks from the upland side during reservoir drawdown;
- scouring of the base of slopes by waves and currents.

As a result of the impact of these processes, existing landslides are activated and new landslides are created, sometimes reaching huge sizes.

However, according to available calculations, large reservoirs on plain rivers can only silt up within a few centuries; reservoirs created on mountain rivers, if appropriate measures are not taken in time, can be filled with sediment within a few decades, and sometimes even faster. Siltation of small mountain reservoirs is controlled by hydraulic flushing; in some cases it is necessary to use mechanical means to clean the reservoirs of sediments.

Sediment deposition in reservoirs changes the regime of solid runoff in the lower reaches. The clarified water intensively erodes the bottom and banks of the river in the lower reaches, thereby causing significant reformation of the river channel. As the water becomes saturated with sediment, the scouring capacity of the river decreases and, at a certain distance from the dam, equalizes with domestic values.

The underwater relief of reservoirs is formed mainly due to bottom leveling by waves and siltation of low areas by wear and tear coming into the reservoir in suspension from the main river and its tributaries and as a result of bank recycling. The process of formation of submarine shoals along the banks has a seasonal rhythmicity. Bank collapse is most intense in spring, and growth and flattening of coastal banks in summer. These processes occur differently in different parts of reservoirs. According to the character of channel regime it is proposed to divide large river reservoirs into five zones:

1) deep-water (dammed) zone. This zone is characterized by the highest water level rise. As a rule, the wave does not deform the bottom. The shores and coastal zone undergo the greatest reformation here. Coastal reworking is the source of a large amount of sediment that forms shoals, spits and bars;

2) the zone of middle depths, which under the NPP differs little from the first zone. At drawdown the reservoir in this zone remains lake-shaped, but becomes shallow; in a number of sections the wave actively interacts with the bottom and levels it; channels and channels are drifted. Coastal reworking is slower than in the lower zone, since scouring almost stops during reservoir drawdown;

3) the upper zone, which at the NPA is sometimes a wide but shallow water body, so the excitement is weakly developed and bank reworking is slow. At reservoir drawdown the mirror narrows to the limits of the domestic channel. The process of wave leveling of the bottom is quite active;

4) the zone of backwater eclipse, which may be a delta or a river, depending on the height of standing levels. Bank reworking is even less significant. Sediment of erosional origin is brought by the main river and lateral outflows. This zone is practically absent in reservoir cascades;

5) a zone of small bays in which no sediment is deposited both from the side of the reservoir, forming a bar, and from above (cone of outflow). A lagoon is created between them. In small bays, the lagoon formed is gradually drifted and the cone of outflow is connected to the bar. There is little overworking of the shore.

These changes in the hydrological regime and other natural processes resulting from the creation of reservoirs and their flow regulation have a significant impact on the water sector. The national economic significance and consequences of the river regime transformation for water sectors are covered in the second section of the book.

Chapter 2. Study and generalization of experience in operation of channel reservoirs on the example of the Orto-Tokoi hydroscheme

2.1 Hydrological and water regime of the reservoir

Artificial obstacles include various hydraulic structures. On rivers, such structures include dams and weirs. A dam blocks the entire river bed and its valley and creates an artificial body of water called a reservoir. Reservoirs are created not only to store water, but also to regulate the flow, i.e. the movement of water in the river.

Flow is characterized by the amount of water flowing over a period of time, e.g. a year, a day. The dam of a large reservoir is a complex hydraulic unit, which combines various hydraulic structures: a hydroelectric power plant (HPP), a spillway for emptying the reservoir, a spillway for discharging water to avoid overfilling the reservoir, an outlet for regulated water deliveries (releases) through the dam, and a fish passage device. As a result of interaction of water masses of the reservoir with its banks, siltation and re-shaping of the banks occurs. Siltation of the reservoir is the process of sedimentation of the reservoir bowl, and bank reshaping is the change in the original shape of the shore slopes due to the destruction of the overwater part of the slopes by waves and the formation of shoals. The water area adjacent to the dam from above and below is called the embankment; above the dam it is called the upper embankment and below the dam it is called the lower embankment. The water level formed in the reservoir as a result of backwatering, i.e. raising the water level by means of the dam, is called the backwater level.

The normal headwater level (NHL) is the highest headwater level that can be maintained under normal reservoir operating conditions. The NPL is used as the basis for reservoir navigation charts. The dead volume level (DWL) is the maximum permissible level of reservoir drawdown (emptying) during reservoir overflow, during the growing season for irrigation of fields.

Whereas in natural inland reservoirs the load balance is dominated by material from surface runoff, in most flatland reservoirs in the first years of their existence the bottom sediments are formed mainly from the products of bank and bed failure; over time their quantity decreases and the load from surface runoff begins to predominate.

In mountain reservoirs with rocky shores, this ratio is significantly different; for example, in the Ortokoy reservoir, the average annual soil input from shoreline abrasion is only 9.1%, while 89.8% comes from river runoff.

The sediment entering the reservoir is distributed over the reservoir bed and forms a ground sediment complex. The composition and distribution of bottom sediments in reservoirs can be very diverse and depends on the size of the drawdown and the degree of flow of the reservoir, the regime of currents and waves, and the quantity and coarseness of the sediment. These factors are strongly influenced by the geographical location of the reservoir, its size and morphology. In each particular reservoir, the nature and quantity of the soil-forming material and the hydrodynamic activity of the water masses are determining factors.

In arid zone reservoirs located in sandy deserts, gross transport of sandy and dusty sediments can be an important source of sediments.

In channel reservoirs the prevailing form of hydrodynamic activity is currents, in valley and lake type reservoirs - waves.

Figure 1. View of the tailrace of the reservoir

In the study of reservoirs, bottom soils were categorized into three main groups:

(a) Primary flooded soils that have not significantly changed their properties;

6) transformed flooded soils that have largely lost their former properties;

7) secondary soils (gravels, sands, silts, the formation of which is preceded by pre-sorting of their particles, and deposits from macroliths).

Figure 2: Middle part, left bank of the reservoir

Figure 3: Middle part, right bank of the reservoir

Pebbles, gravelly and coarse-grained sands are developed mainly in mountainous and foothill reservoirs, mainly in the deltas of rivers flowing into the reservoir. Sands by formation can be divided into channel, coastal and open space sands, silty sands are formed at the boundary of sand distribution, where changes in hydrodynamic activity are observed. Sandy and gray silts are divided by their origin into alluvial and local silts; the former are formed from alluvial sediments brought into the reservoir, the place of their accumulation being the zones of backwater wedging, while local silts are formed mainly from the products of erosion of the shore and reservoir bed. Remains of peat, submerged terrestrial vegetation, products of wind transport, and remains of aquatic fauna and flora are involved in the formation of all sediments.

Depending on the hydrodynamic activity (currents, waves), the depth of the sands at these reservoirs varies.

The process of underwater relief reorganization starts simultaneously with reservoir filling. Scouring and re-deepening of the bottom by currents is more often observed in channel reservoirs. Scouring as a result of wave activity usually occurs in water areas with depths not more than 8-10m. Bottom leveling, both in shallow and deep water zones, occurs due to sediment deposition in relief depressions.

Suspended material from the agitation zone enters the open part of the water body, where it is deposited on the bottom, forming a layer of sand-silt or silt sediments. Their greatest thickness is observed in closed depressions (flooded lakes, oxbows and interdrift depressions). The thickness of sediments decreases with distance from the shore and towards the dam.

The process of reservoir bed formation, as well as the process of bank formation, is divided into two periods:

- submarine landform formation, accompanied by more or less sediment input, and submarine landform stabilization, which occurs when sediment input to the reservoir is moderate;

In the second stage, shallow shores also intensively collapse and the main elements of the bottom relief are formed; in shallow water areas, wave leveling of the bottom begins, and in the deep-water zone - primary relief leveling. The duration of this stage may exceed 2-4 decades.

The second period is characterized by a marked decrease in the input of material from abrasion and an increasing role for solid runoff from the catchment and the products of aquatic life. During the first stage of this period, the processes of sediment transport along the shore and the detachment and filling of bays play a leading role in the formation of the reservoir bed; there is also a slow overgrowth of shallow waters and the formation of a complex of bottom sediments is completed. During the second stage, the relief of the reservoir bed is formed by the accumulation of accumulative material further and further from the shore, which contributes to the gradual detachment of some areas of the reservoir and the leveling of the bottom relief. The duration of this stage may reach several hundred years.

Significant changes in the formation of bottom sediments and, consequently, bed relief occur in case of reservoirs' arrangement in a cascade. In case of continuous arrangement of reservoirs, the amount of solid runoff decreases noticeably towards the river mouth. The main receiver of solid river runoff is the upper reservoir, where it is accumulated. All downstream reservoirs receive suspended sediment only during the transit discharge of spring floods through the cascade and from tributaries. In this case, the inflow of solid runoff into the river delta is drastically reduced.

By accumulating the vast majority of sediment, reservoirs, and even more so reservoir cascades, drastically reduce the amount of sediment entering deltas or estuaries.

The creation of reservoirs changes the regime of sediment movement. The nature and extent of these changes depend on many factors: the size of the reservoir, its outline in plan, the size of the drawdown of the reservoir, the quantity and coarseness of the sediment carried by the river, the extent of bank recycling and the flow capacity of the reservoir, i.e. the exchangeability of the water in the reservoir, expressed as the ratio of the average annual flow to the total volume of the reservoir

Due to the dramatic reduction in flow velocity, concentrated deposition of sediment, usually 0.25mm and above, occurs in the tailrace of the reservoir, similar to that observed in river mouths. The finer sediment is deposited throughout the reservoirs and is partly carried downstream. When the reservoirs are drawn down, the point where the backwater curve wedges out moves in the direction of the dam, so that the previously deposited sediment is eroded and moved in the same direction, and when the reservoir is filled the point of deposition moves upriver again. The periodic movement of the site of concentrated sediment deposition leads to the filling of the Dead Volume of the reservoir. The reworking of the banks and alongshore sediment transport also contributes to this.

In some cases, landslides and bank collapse have a significant impact on the sedimentation and siltation of mountain reservoirs. Creation of deep reservoirs with large (up to 50-80 m) and comparatively fast level drawdown leads to disturbance of stability of their steep height banks. These violations consist of the following:

- weighing of rocks during filling and their weighting (due to high water saturation) during discharging;
- weakening of rock strength and reduction of its shear resistance as a result of variable water saturation and drying, weathering, leaching, etc;
- development of hydrodynamics of water pressure on rocks from the uphill side during reservoir drawdown;
- scouring of the base of slopes by waves and currents.

As a result of the impact of these processes, existing landslides are activated and new landslides are created, sometimes reaching huge sizes.

However, according to available calculations, large reservoirs on plain rivers can only silt up within a few centuries; reservoirs created on mountain rivers, if appropriate measures are not taken in time, can be filled with sediment within a few decades, and sometimes even faster. Siltation of small mountain reservoirs is controlled by hydraulic flushing; in some cases it is necessary to use mechanical means to clean the reservoirs of sediments.

Sediment deposition in reservoirs changes the regime of solid runoff in the lower reaches. The clarified water intensively erodes the bottom and banks of the river in the lower reaches, thus causing significant reformation of the river channel. As the water becomes saturated with sediment, the scouring capacity of the river decreases and, at a certain distance from the dam, equalizes with domestic values.

The underwater relief of reservoirs is formed mainly due to bottom leveling by waves and siltation of low areas by wear and tear coming into the reservoir in suspension from the main river and its tributaries and as a result of bank recycling. The process of formation of submarine shoals along the banks has a seasonal rhythmicity. Bank collapse is most intense in spring, and growth and flattening of coastal banks in summer. These processes occur differently in different parts of reservoirs. According to the character of channel regime it is proposed to divide large river reservoirs into five zones:

1) deep-water (dammed) zone. This zone is characterized by the highest water level rise. As a rule, the wave does not deform the bottom. The shores and coastal zone undergo the greatest reformation here. Coastal reworking is the source of a large amount of sediment that forms shoals, spits and bars;

2) the zone of middle depths, which under the NPP differs little from the first zone. At drawdown the reservoir in this zone remains lake-shaped, but becomes shallow; in a number of sections the wave actively interacts with the bottom and levels it; channels and channels are drifted. Coastal reworking is slower than in the lower zone, since scouring almost stops during reservoir drawdown;

3) the upper zone, which at the NPA is sometimes a wide but shallow water body, so the excitement develops poorly and bank reworking is slow. At reservoir drawdown the mirror narrows to the limits of the domestic channel. The process of wave leveling of the bottom is quite active;

4) a zone of backwater eclipse, which may be a delta or a river, depending on the height of standing levels. Bank reworking is even less significant. Sediment of erosional origin is brought by the main river and lateral outflows. This zone is practically absent in reservoir cascades;

5) a zone of small bays in which no sediment is deposited both from the side of the reservoir, forming a bar, and from above (cone of outflow). A lagoon is created between them. In small bays, the lagoon formed is gradually drifted and the cone of outflow is connected to the bar. There is little overworking of the shore.

These changes in the hydrological regime and other natural processes resulting from the creation of reservoirs and their flow regulation have a significant impact on the water sector. The national economic significance and consequences of river regime transformation for the water sectors are covered in the second section of the book.

2.2 Channel deformation in the upper reaches of reservoirs

Currents on reservoirs arise under the action of wind and water runoff. Wind currents are observed far from the shores and their velocities are relatively low. Runoff currents depend on the reservoir site and the growing season. In the lower part of the reservoir, flow can be quite rapid during spring floods and dam releases. During floods, the flow velocity in the narrow parts of the reservoir can be comparable to that of rivers.

Attempts to characterize the flow in the reservoir bowl did not give the desired complete and detailed results due to the imperfection of the float method of recording the direction and values of velocities under conditions of constant water surface excitement and winds reaching an average of 3 points.

The floats used, as well as the special Alexeev's swivel system, distorted the idea of the actual picture of currents underneath the river

by even the slightest wind. Therefore, these works were carried out at a time when there was a change in wind direction.

The character of currents was determined at three reservoir fills.

The data on incoming water and discharge rates and lengths of backwater curves in the Orto-Tokoi Reservoir during the survey are given in Table 1.

At water mark in the reservoir 727.3m, after three times filling and discharging of the above mark, the flow velocities at the section of the backwater curve wedging out (at a distance of 5000m from the dam) amounted to 1.4 -1.1m/sec against the domestic river velocities of 2.4 -2.7m/sec. Downstream of the initial station at 330m the velocities decrease to 0.8m/sec, and even lower at 260m become equal to 0.6m/sec. It should be noted that higher values of velocity in the considered stations are observed in the sections of flooded riverbed, and lower values in the sections of flooded floodplain.

Lengths of backstop curves

Table 2.1.

year	Date of filming production	Reservoir water horizon	Inlet flow rate, m^3/sec	Discharge from the reservoir, m^3/sec	Length of the backwater curve, m

2006	25.07 -10.07	727,3	55,0	60,0	5020
2007	19.07 -25.07	741,5	48,5	71,0	10600
2008	6.07 -12.07	748,8	30,0	115,0	12200

This pattern of velocity distribution is observed at an in-stream flow of 55.0m^3/sec and a discharge of 60.0m^3/sec.

At water level in the reservoir 741.5m, the flow velocity at the section of the backwater curve wedging out (above the dam site 10600m) and below 120m (site№ 8) decreases to 2.0m/sec against the domestic river flow velocity of 2.8-3.0m/sec.

Below gauge 8 at 700 meters velocities do not exceed 0.35 -0.06m/sec and in gauge 6 decrease to 0.045 -0.06m/sec.

Similar dynamics of velocity distribution in the zone of the backwater curve wedging out is observed when the reservoir is filled up to the mark 748.8. While flow velocities in domestic conditions fluctuate within 2.5 -2.8m/sec (site 12), downstream at 1200m (site 11) velocities decrease to 1.4 - 1.6m/sec and in the zone of the backwater curve wedging out between sites 9 and 10 in the distance from the dam 12200m decrease to 0.6 -0.8m/sec, in the immediate vicinity of site 9 velocities reach 0.1 -0.08m/sec.

It should be noted that current velocities and directions were measured with the engineer Alexeev's rope. However, due to the labor-intensive use of this rope and the imperfection of swimming facilities, the above-mentioned work was carried out only between gauging stations 11 and 9.

When the reservoir is filled to 727.3 (dead volume) and the transit discharge is 55m^3/sec, the velocities increase noticeably only in the immediate vicinity of the dam. Thus, if at site 4 (4000m from the dam) velocities are equal to 0.008m/sec, at a distance of 300m from the dam their value is 0.2 - 0.3m/sec, at a distance of 60m - 0.5m/sec and just before the operational tunnel 1.3 - 1.8m/sec.

From the results of the analysis of data on the reservoir velocity regime, it can be concluded that the highest flow velocities in different reservoir gauges are observed above the flooded domestic channel.

This character of currents indicates the presence of a large capacity of stagnant zones (in the flooded floodplain) along the length of the reservoir, which obviously create favorable conditions for the settling of the smallest particles of suspended sediment.

This is confirmed by the large thickness of the total siltation layer in the stagnant zones. It should be noted that the deposition of suspended sediment occurs at a fairly short distance from the zone where the backwater curve wedges out. At a river discharge of 48.5m^3/sec (20 July) and with the reservoir filled to 741.8m, the turbidity distribution was as follows:

Table 2.2.

Turbidity in the wedging zone

gauges	8	8 - 270	8 - 800	8 - 950	8 - 1200	7
turbidity, gr.liter.	0,28	0,24	0,22	0,015	0,013	clean water

Similar deposition of suspended sediment can be observed at other reservoir fills. Almost complete deposition of suspended sediment occurs at a distance of 800-1200m from the zone where the backwater curve wedges out (within reaches 5-12) at flow velocities of about 0.2m/sec and less.

The uneven distribution of suspended sediment over the length and width of the reservoir is also explained by the sharp fluctuations in the water horizons of the reservoir and the reservoir when the reservoir is drawn down, part of the suspended sediment is carried downstream and deposited in areas with lower water velocities.

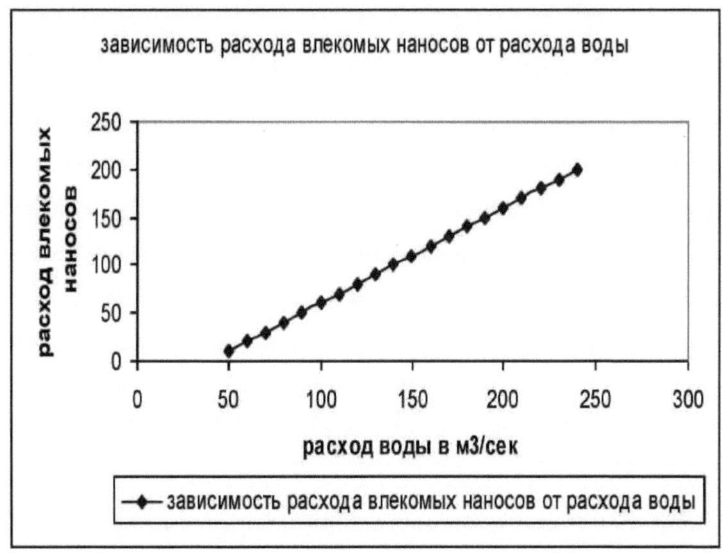

Figure 4: Dependence of transported sediment discharge on water flow rate

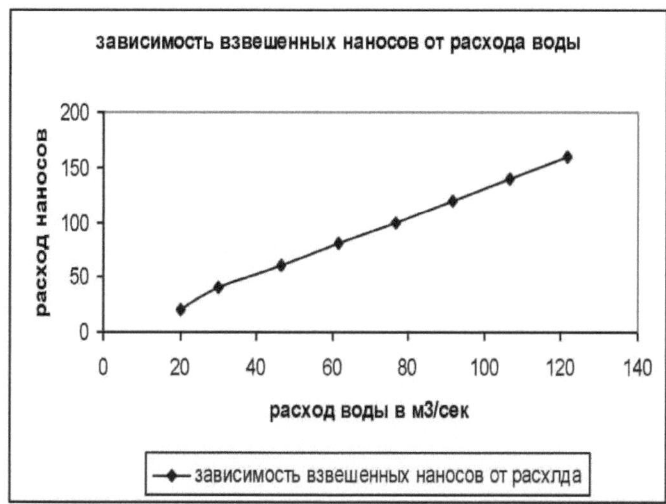

Figure 5: Suspended sediments as a function of flow rate

2.3 Channel deformation in the downstream channel and changes in natural conditions in areas adjacent to the reservoir at the hydroelectric facility

Changes in the landscape and transformation of the river regime by river valley reservoirs have a significant impact on changes in the natural environment of adjacent areas. The impact of reservoirs on natural conditions is extremely diverse. It can be manifested directly and indirectly, can be positive and negative permanent and temporary.

The creation of reservoirs causes a number of changes in the natural conditions and economy of the affected area:

- flooding;
- flooding and erosion of lands used both as agricultural lands and those on which settlements, railroads and highways are located, which leads to the disruption of established economic and other links;
- increase of groundwater level in the adjacent territory;
- waterlogging of lands and changes in soil and vegetation cover;
- changes in climatic conditions (above the reservoir and on the banks) and sanitary-hygienic conditions in the reservoir area.

Within the upper reservoir, these changes are most pronounced in the lower part of the reservoir, where the head, water mass, and mirror width reach the highest values for the reservoir, and least pronounced in the zone of head wedging, where changes compared to natural conditions are less significant or not traceable at all.

Creation of reservoirs that change the river regime causes disturbance of natural conditions and economic environment also in the river valley downstream of the hydroscheme. These changes are manifested to a greater or lesser extent over a large length. The impact of regulation of river flow of the reservoir is also reflected on the nature and economy of the estuaries of seas and lakes, into which the regulated rivers flow. When a river is backed up, its water mirror expands considerably.

Reservoir bank reformation is understood as the process of forming a new bank as a result of erosion, collapse, landslide, rockfall, scree, subsidence and other relief deformations in the zone of a new water cut, as well as deposition of eroded and collapsed rocks and sediment brought by the river and its tributaries. The extent of bank reformation is very significant; abrasion covers more than 50% of the length of the banks of many reservoirs. Where the reservoir banks are composed of sand or loess, each storm in the first years after the reservoir is filled leads to bank retreat from 1-5 to 10-15 meters. The nature and extent of bank reformation is influenced by a number of factors:

- hydrological regime of the reservoir;
- geologic and geomorphologic factors;
- human economic activity.

The release of water from the reservoir unloaded with sediment results in a general deformation of the river channel in the lower reach. Changes in the position of the average river bed bottom in the lower reach were determined by vertical survey of 25 cross sections in a 29.4 km long section from the tunnel outlet portal.

The data of scour and debris volume calculation by stations on the vertical survey of the Orto-Tokoy reservoir are presented in Tables 2.3, 2.4, 2.5.

Scour and channel debris volumes in the downstream channel of the Orto-Tokoy dam for the period of 2007.

Table 2.3.

No. of trunk	Dam growth M	average depth		Area		Average area		Flush volumes THOUSAND. M	The volume of the blockage z THOUSAND. M	difference z THOUSAND. M
		scour, M	rubble, M	scour M^2	rubble M^2	scour M^2	rubble M^2			

1	2	3	4	5	6	7	8	9	10	И
13 14-15	1590 2180	0,00	0,00	0,00	0,00	1,45	0,55	1,02	0,38	-0,64
16	2880	0,08	0,03	2,90	1,10	3,45	0,55	4,47	0,71	-3,76
17	4180	0,10	0,00	4,80	0,00	4,40	0,00	9,70	0,00	-9,70
18	6380	0,085	0,00	4,10	0,00	3,50	4,35	9,10	11,30	2,20
19	8980	0,08	0,22	3,00	8,70	2,50	4,45	7,00	11,70	4,70
20	11680	0,08	0,005	2,10	0,20	6,95	0,35	13,90	0,70	-13,20
21	13680	0,15	0,003	11,80	0,50	7,00	11,50	15,40	25,30	9,90
22	15880	0,02	0,17	2,20	22,50	2,60	11,25	6,50	28,10	21,60
23	18380	0,08	0,00	3,00	0,00	2,70	0,00	9,40	0,00	-9,40
24	21880	0,065	0,00	2,50	0,00	2,65	0,00	10,40	0,00	-10,40
25	25880	0,02	0,00	2,80	0,00	1,40	0,00	4,90	0,00	-4,90
26	29380 Total				0,00			91,8	78,1	-13,7

- washout + rubble

Scour and channel debris in the downstream channel of the Orto-Tokoy dam for the period 2007 - 2008.

Table 2.4.

No. of trunks	Dam growth M	average depth		Area		Average area		Flush volume	Debris volume	difference
		erosions	rubble	erosions	rubble	erosions	rubble			
		M	M	M^2	M^2	M^2	M^2	3 THOUSANDS. M	3 THOUSANDS. M	3 THOUSANDS. M
1	2	3	4	5	6	7	8	9	10	И
4	100	1,16		40,00		19,75		0,99		-0,99
5	150	0,64		19,50		10,225		0,51		-0,51
6	200	0,05		1,30		5,00		0,50		-0,50
7	300	0,31		8,70		4,30		0,43		-0,43

8	400					1,75		0,18		-0,18
9	500	0,17		3,50		2,575		0,52		-0,52
10	700	0,066		1,65		1,15		0,59		-0,59
12	1300	0,02		0,65		0,33		0,07		-0,07
13	1590			0,00		1,33		0,78		-0,78
14-15	2180	0,087		2,60		3,15		2,21		-2,21
16	2880	0,10		3,70		2,15		2,80		-2,80
17	4180	0,01		0,60	0,00	0,30	1,05	0,66	2,30	1,64
18	6890		0,04	0,00	2,10	0,00	3,05	0,00	7,90	7,90
19	8980		0,01	0,00	4,00	0,05	2,70	0,14	7,30	7,16
20	11680	0,002	0,030	0,10	0,15	0,05	3,20	0,10	6,40	6,30
21	13680		0,034	0,00	5,00	1,125	2,50	2,48	5,50	3,02
22	15880	0,002		2,25	0,00	1,275	0,00	3,19	0,00	-3,19
23	18380	0,007		0,30	0,00	0,15	1,60	0,53	5,60	5,07
24	21880		0,08	0,00	3,20	0,425	2,05	1,70	8,20	6,50
25	25880	0,007	0,007	0,85	0,90	0,425	2,05	1,70	8,20	6,50
26	29330 Total					0,425	0,45	1,5 19,9	1,5 44,7	0,00 24,8

Scour and channel debris volumes in the downstream channel of the Orto-Tokoy dam for the period 2008 -2009.

Table 2.5.

No. of trunk	Dam growth M	average depth		Area		Average area		Flush volume	Debris volume	difference
		erosions	rubble	erosions	rubble	erosions	rubble			
		M	M	M^2	M^2	M^2	M^2	3 THOUSANDS. M	3 THOUSANDS. M	3 THOUSANDS. M
1	2	3	4	5	6	7	8	9	10	И
4	100	0,04		1,40		1,85		0,09		-0,09
5	150	0,07		2,30		1,150		0,06		-0,06
6	200			0,00		1,20		0,12		-0,12
7	300	0,09		2,40		4,00		0,40		-0,40
8	400	0,35		5,60		4,00		0,40		-0,40
9	500	0,12		2,40		1,20		0,24		-0,24

10	700			0,00		1,45		1,27		-1,27
13	1590	0,63		2,90		2,65		1,59		-1,59
14-15	2180	0,080		2,40		2,40	0,55	1,68	0,38	-1,30
16	2880	0,07	0,03	2,40	1,10	1,25	1,35	1,62	1,75	0,13
17	4180	0,00	0,03	0,10	1,60	0,05	1,60	oh, and	3,50	3,39
18	6380		0,001	0,00	0,60	3,10	2,70	8,00	7,00	-1,00
19	8980	0,15	0,07	6,20	2,80	3,10	1,40	8,40	3,70	-4,70
20	11680			0,000	0,00	0,80	0,50	1,60	1,00	-0,60
21	13680	0,01	0,007	1,55	1,00	2,200	1,20	4,85	2,60	-2,25
22	15880	0,020	0,01	2,90	1,50	1,450	0,75	3,60	1,87	-1,73
23	18380			0,00	0,00	1,00	0,25	3,50	0,88	-2,62
24	21880	0,024	0,013	2,00	0,50	1,000	1,25	4,00	5,00	1,00
25	25880		0,016	0,00	2,00	0,800	1,00	2,80	3,50	0,70
26	29330	0,04		1,60	0,00					0,00
	Total							44,33	31,18	-13,15

Only the section between stations 19 and 26 at 9km and 29km was found to be deformed. Table№ 2.3 shows that the total volume of scour was 91.8tm^3 and debris 78.1tm^3. The greatest bottom depression was observed between gauges 19 and 23 and is 0.1 - 0.15m, in other gauges on the contrary, the riverbed is overburdened, which is most pronounced in gauges 19 - 22, where the average height of the overburden reaches 0.17 - 0.22m.

Further deformation of the river channel is shown in Table No. 2.4.

The total volume of riverbed debris was 44.7tm^3. The volume of scour is only 19.9tm^3. The largest volume of scour was observed in the section 13 - 19 and 21 - 23 gauges and the average depth of scour reaches about 0.1m, while a significant section of the river from 17 to 25 gauges was subjected to debris and the average height of the debris was also 0.1m.

During the period 2007 - 2008 significant deformation occurred in the section between gauges 13 - 24, where the average depth of channel scour between gauges 18 - 24 is 0.1m and the average height of debris in gauges 16 - 17 does not exceed 0.05m. The total volume of scour amounted to 44.3tm$^{(3)}$and 31.2tm^3, i.e. the channel sediments were washed away.

More significant deformation of the river channel has occurred in the section between stations 16 and 26, with intensive scour between stations 16 and 20 and debris between stations 24 and 26.

The average depth of scour is 0.1m, and the average height of the blockage is 0.08m, the total volume of scour is 147.34tm^3 and blockage is 141.04tm^3. These deformations of the river channel occur under conditions of different duration and size of discharge discharges from the reservoir.

Analysis of research materials and observational data allow us to draw the following conclusions:

- average monthly temperatures of the Orto-Tokoi MS exceed on average the data of the Kochkor MS by +4^{0}C, while the amount of precipitation in the reservoir area is 2.7 times less than in the Kochkor valley.

Plan of the Chu River channel downstream and upstream of the Orto-Tokoy dam

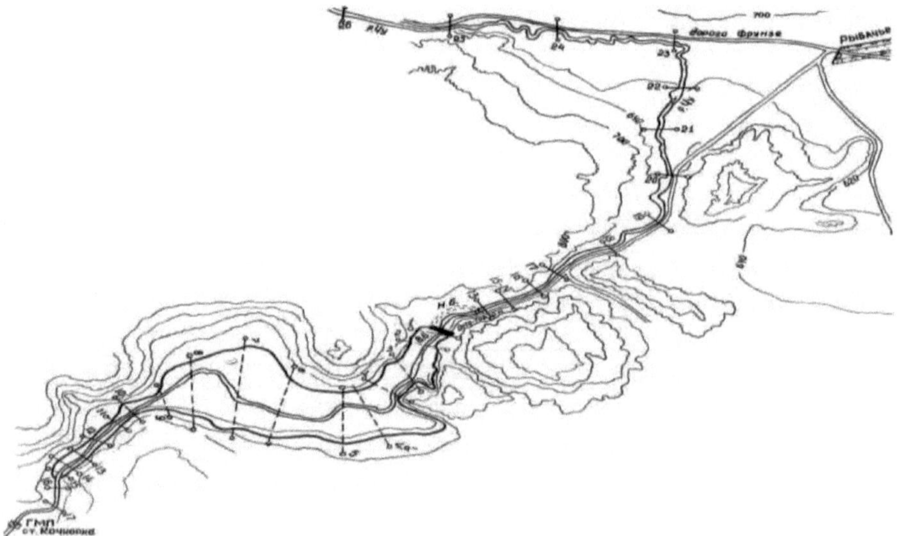

Figure 6. Plan of the Chu River channel downstream and upstream of the Orto-Tokoy dam

- The maximum average monthly air temperatures occur in May-August and reach +27, +31^{0}C.
- Under water-balance calculations, the average annual precipitation from the area enclosed by the Kochkor post and dam should be taken as 5mn m^3 and evaporation from the water surface of the Orto-Tokoy reservoir as 17mn m^3;
- calculation of evaporation from the reservoir water surface according to meteorological observations (moisture deficit), should be performed using the dependence $E = 0.65\Sigma d$, which gives the results that most closely correspond to in-situ observations;
- The highest flow velocities along the length of the reservoir at different fills are observed above the flooded domestic channel. Favorable conditions for the settling of suspended sediment are created in the flooded floodplain zones;

- The annual runoff of sediment and suspended sediment averages about 244,000 m^3 or about 1/88 of the dead volume of the reservoir;

- The release of water from the reservoir unloaded with sediment causes general deformation of the diversion channel. The most intensive deformation occurred in the section between sites 13-26, distant from the dam at distances of 1580 and 29380, respectively;

- The average annual volume of scour due to self-paving of the channel bed is small and leaves about 6 - 13 thousand meters ^{3};

- scour occurs due to lowering of the river bed and bank deformation;

- as a result of washing out of fine particles composing the channel, the average diameter of sediments increases by 10 -20% per year;

- Such insignificant general deformations of the diversion channel indicate that dangerous river bed elevations directly behind the tunnel outlet portal and in the mouth zone of the catastrophic discharge and even at large water releases from the reservoir should not be feared.

2.4 Reservoir operation for the period 2004-2007

Filling of the reservoir to accumulate water for irrigation was carried out from April 13 to May 31, when 65mn^3 or 14.5% of the design volume was accumulated. The maximum water level upstream of the dam reached 738.2m and the reservoir was emptied from April 27 to the end of July 2006, after which the incoming flow into the reservoir was discharged downstream through the construction tunnel portal.

The 2006 filling was accomplished by closing the portal opening. Opening of the portal opening was performed as water horizons were depleted and discharge flow rate increased.

Accumulation in 2007 was carried out from February 28 to May 16. Water horizon elevation reached 739.6m. Water discharge to the lower reach was from 20 to 80m^3/sec (August). During the flood period (by June 14), the reservoir filling reached its maximum (mark 740) with a volume of 84.4 million m3. The method of closing and opening the construction tunnel portal opening was similar to the previous one.

In a short period (13 days) from a high of 744.2 to a low of 720.9 (May 6).

Emptying the reservoir

Date	Water discharge in m^3/sec	Date	Water discharge in m^3/sec
17.04	110,0	5.06.	43,0

18.04	156,0	7.06	0,6
21.04	138,0	3.06	3,5
30.04	103,0	4.06	15,0

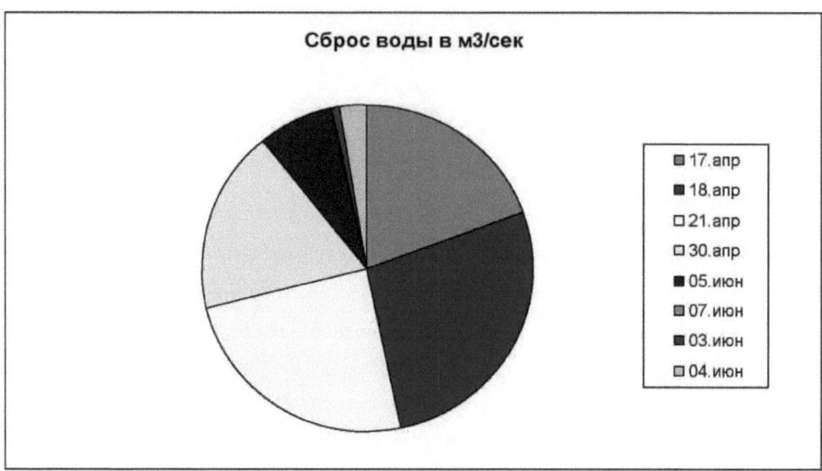

Figure 15.

When the reservoir was filled, the water horizon reached the level of 745.3m. Water was released from the reservoir according to the water use plan.

The maximum water level of the reservoir in 2007 at 747.3 and 177 mln m^3 refers to June 7. The reservoir was released by manipulating first the emergency and later the cone gates of the production tunnel. H

Accumulations in the reservoir during the whole operation period continue from October to March up to the mark of 752.2m and the volume up to 273, 2 million m3. From the beginning of March to June water from the reservoir is discharged to the lower volume. During floods, releases are insignificant (30-70m^3/sec), so the reservoir is subsequently filled, and the maximum level reaches 754.1m (mid-June) and the volume of water is 311.2 million m3. Drawdown of accumulated volumes continues until September 19, after which filling begins again. By January 2006, the volume of water in the reservoir reached 268.0 million m3 at a water level of 751.97m.

In order to maintain these water horizons and ensure operation of the HPP, from November to May 2007, partial drawdown of the reservoir occurred at flow rates of 10 - 18m^3/sec.

Starting from May 13, 2008, due to the dry summer intensive water release from the reservoir (35 -

115 m^3/sec) was carried out. If on May 13 the maximum water horizon of the reservoir was 754.7m with a volume of 314.0 million m3, then by September the water horizon dropped to 728.5 with a volume of water in the reservoir of only 11.4 million m3. Subsequent reservoir storage begins in October and will continue through March 13. The release from the reservoir from March 13 to June 6 will be 3.5m^3/sec, so was filling the reservoir.

The maximum water level in 2005 was on June 14 and reached 756.75m, the accumulated volume in the reservoir was 356.65 million m3. The reservoir was emptied to the dead volume horizon in a short period of time from June 15 to September 15, after which the reservoir began to fill, reaching a volume of 210 million ^{m3} by January 1 at a water table elevation of 749.01m. Data on the dynamics of reservoir filling and emptying for the period 2004-2007 are shown in Figures 16, 17.

Figure 16. Reservoir filling dynamics.

Figure 17. Emptying of the reservoir.

2.5. River flow and water flow patterns

The flow of water in rivers is caused by the gradient of the water surface from source to mouth: the greater the gradient, the higher the flow velocity. This velocity also depends on the average depth, sinuosity and roughness of the channel. It is different on the spillways and rapids: in low water, it is greater on the rapids, and in high water - on the spillways. Flow velocity decreases from the source to the mouth of the river. The flow velocity of a river is the average flow velocity of individual jets of the river flow. The velocities of jets are not the same at different points in the cross section of the stream, the area of which is called the live section of the river channel. The jet velocity is maximum over the deepest point near the water surface at a point about 0.2 depth from the surface. From this point, the velocity of the jets decreases toward the bottom of the river and its banks. At the water surface, the velocity of the jets is maximum above the deepest point. The line connecting the points on the water surface with the highest jet velocities and the greatest depth of the channel is called the river stem. At the bends of the channel simultaneously with the flow along it is observed and transverse flow: at the surface - towards the concave bank, at the bottom - towards the convex. So on curvilinear sections of the river, the water flow seems to "screw" along the channel downstream. This screw-shaped water flow "carries" the river stem closer to the concave bank. Thus, on rectilinear sections of the river the flow velocity is highest in the middle of the channel, on curvilinear sections - closer to the concave bank. When the water level changes, two additional transverse currents appear between the middle of the river and its banks. When the water rises, these currents on the water surface are directed from the middle to the banks, on the bottom - from the banks to the middle, and in the middle of the river - from bottom to top. When the water recedes, it is the other way around. All rivers are affected by sediment deposition. The most characteristic type of sediment deposition is the spit, which forms at the convex bank in the form of a wedge running at an angle downstream. The spit gradually goes under water, extending far into the river channel. Sediment deposits in the channel can be either underwater or overwater. Rolls. This is a stable sediment deposit in the form of a berm crossing the river channel (often called a ford). It is the main factor complicating navigation on rivers. Most often rolls are located in places where the water flow transitions from one bend to another. In high water, flow directions are parallel to the banks of the rolls, in low water - depending on their features. For example, the formation of a swamp current depends on the shape of the basement: for example, on an overkat with a flat basement, the well-defined river stem coincides with the trough axis and there are no swamp currents. Solid material is transported by rivers in three ways: dissolved, suspended and dragged along the bottom. The first method plays almost no role in the formation of placers, the second one plays a very subordinate role (in the case of transportation of gold), and the third one is the main one. The main mass of solid material is carried by the river during floods. At this time the amount of water and the speed of its flow are so great that the river sets in motion a considerable layer of sediment covering its bottom. This layer of moving sediment is usually several decimeters thick and sometimes 1-2 m thick. It consists of pebbles immersed in a liquid mush of water, sand and muddy particles, i.e. a kind of mud flow. In its lower parts it has a thicker consistency, which gradually becomes more liquid in the upper parts of the layer. In the same way, the speed of movement changes: it is minimal (up to zero) in the lower parts of the layer and maximum at the top. The larger the flood, i.e. the greater the water mass and current velocity, the greater the power of the moving bedload layer. The movement of pebbles and larger particles is not so much by rolling as by the sliding of one layer over another. It is quite clear that during this movement valuable metals, as well as increasingly heavier minerals, inevitably find their way to the lowest parts of the moving layer and are concentrated in the slowest moving part of the layer, which is only a few centimetres thick. Thus, already in the very process of solid material transportation, two layers are separated, with and without heavy minerals, which are like prototypes of sands and peats. As soon as the high water subsides, i.e. its mass and flow velocity decrease, the movement of the bottom sediments is immediately weakened. The lowest part of them, the proto-sands, so to speak, becomes immobile, fixed. As the water recedes, the upper parts of the moving layer become fixed and its power decreases. During low water, the movement of the bottom sediment almost ceases and takes place mainly by rolling over the bottom of individual pebbles. In the next flood, the sediment is again in motion. If the flood is of the same or greater magnitude than the previous one, the layer containing the valuable metal will move. If the flood is of lesser magnitude, this layer does not move. Even in this case, however, the lowest part of the moving bedload layer may contain valuable metal, provided that it is brought in from upstream. When the high water recedes, this new metal-bearing layer will also build up on top of the previous one. If a given stretch of stream is in the stage of sediment accumulation, i.e. the amount of material brought in is greater than that taken out and the total thickness of the sediment is gradually increasing, the metal-bearing layer will build up from year to year by fixing the lowest part of the moving bottom sediment. Accumulating over the course of centuries, this accumulation leads to the formation of a more or less thick layer of metal-bearing sands. The accumulation of sands lasts until the influx of valuable metal from upstream parts of the river stops. After this, peat, i.e. the same sediment but without the valuable metal content, accumulates in the same way. If all floods were of exactly the same magnitude, then each year an equal, very small part of the bedload layer would be fixed, corresponding to the general rate of accumulation of sediment in a given section of the river. But there are usually alternating high and low floods, and even whole periods of both. Exceptionally high water can draw a long-established bedload layer into motion, wash over it and redistribute valuable metal within it. If such a high flood does not occur again, or occurs after a long interval when the total thickness of the sediment has increased considerably, this layer will remain permanently immobile. The layer from the next exceptionally high flood will be deposited on top of it in the same way. Thus the river load that makes up the alluvial placers is not the sediment of intermediate or even high waters, but only of the exceptionally high floods experienced by the river during the period of accumulation. This is the principle of the gradual accumulation of river-bottom contributions.

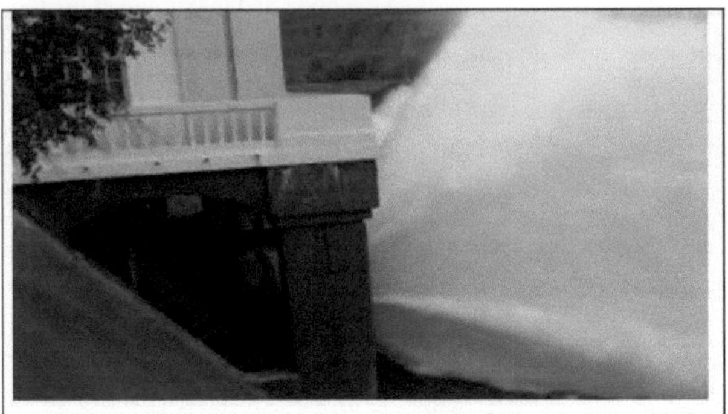

Figure 18. View from downstream, water discharge

Figure 19. View from downstream, conical gates.

2.6. Channel deformation in upstream and downstream reservoirs

Attempts to characterize the flow in the reservoir bowl did not give the desired complete and detailed results due to the imperfection of the float method of recording the direction and values of velocities under conditions of constant water surface excitement and winds, reaching an average of 3 points.

The floats used, as well as a special Alexeev's swivel system distorted the idea of the actual picture of currents under the influence of even insignificant winds. Therefore, these works were carried out during the period of time when a change of wind direction was observed.

The character of currents was determined at three reservoir fills.

The reservoir inflow and discharge rates and the lengths of the backwater curves during the survey are summarized in Table 10.

Lengths of backstop curves

year	Date of filming production	Reservoir water horizon	Inlet flow rate, m^3/sec	Discharge from the reservoir, m^3/sec	Length of the backwater curve, m
2006	25.07 -10.07	727,3	55,0	60,0	5020
2007	19.07 -25.07	741,5	48,5	71,0	10600
2008	6.07 -12.07	748,8	30,0	115,0	12200

At water mark in the reservoir 727.3m, after three times filling and discharging of the above mark, flow velocities at the section of the backwater curve wedging out (at a distance of 5000m from the dam) amounted to 1.4 -1.1m/sec against domestic river velocities of 2.4 -2.7m/sec. Downstream of the initial station at 330m the velocities decrease to 0.8m/sec, and even lower at 260m become equal to 0.6m/sec. It should be noted that higher values of velocity in the considered stations are observed in the sections of flooded riverbed, and lower values in the sections of flooded floodplain.

This pattern of velocity distribution is observed at an in-stream flow of 55.0m^3/sec and a discharge of 60.0m^3/sec.

At water level in the reservoir 741.5m, the flow velocity at the section of the backwater curve wedging out (above the dam site 10600m) and below 120m (site No. 8) decreases to 2.0m/sec against the river's domestic velocity of 2.8-3.0m/sec.

Below gauge 8 at 700 meters velocities do not exceed 0.35 -0.06m/sec and in gauge 6 decrease to 0.045 -0.06m/sec.

Similar dynamics of velocity distribution in the zone of the backwater curve wedging out is observed when the reservoir is filled up to the mark 748.8. While flow velocities in domestic conditions fluctuate within 2.5 -2.8m/sec (station 12), downstream at 1200m (station 11) velocities decrease to 1.4 -1.6m/sec and in the zone of the backwater curve wedging out between stations 9 and 10 at a distance from the dam of 12200m decrease to 0.6 -0.8m/sec, in the immediate vicinity of station 9 velocities reach 0.1 -0.08m/sec.

It should be noted that current velocities and directions were measured with the engineer Alekseev's rope. However, due to the labor-intensive use of this rope, as well as the imperfection of swimming facilities, these works were carried out only between Stations 11 and 9 (Fig. 23).

When the reservoir is filled to 727.3 (dead volume) and the transit discharge is 55m^3/sec, the velocities increase noticeably only in the immediate vicinity of the dam. Thus, if at site 4 (4000m from the dam) velocities are equal to 0.008m/sec, at a distance of 300m from the dam their value is 0.2 - 0.3m/sec, at a distance of 60m - 0.5m/sec and just before the operational tunnel 1.3 - 1.8m/sec.

From the results of the analysis of data on the reservoir velocity regime, it can be concluded that the highest flow velocities in different reservoir gauges are observed above the flooded domestic channel.

This character of currents indicates the presence of a large capacity of stagnant zones (in the flooded floodplain) along the length of the reservoir, which obviously create favorable conditions for the settling of the smallest particles of suspended sediment.

This is confirmed by the large thickness of the total siltation layer in the stagnant zones. It should be noted that the deposition of suspended sediment occurs at a fairly short distance from the zone where the backwater curve wedges out. At a river discharge of 48.5m^3/sec (20 July) and with the reservoir filled to 741.8m, the turbidity distribution was as follows:

Turbidity in the wedging zone

gauges	8	8 - 270	8 - 800	8 - 950	8 - 1200	7
turbidity, gr.liter.	0,28	0,24	0,22	0,015	0,013	clean water

Similar deposition of suspended sediment can be observed at other reservoir fills. Almost complete deposition of suspended sediment occurs at a distance of 800-1200m from the zone where the backwater curve wedges out (within reaches 5-12) at flow velocities of about 0.2m/sec and less.

The uneven distribution of suspended sediment over the length and width of the reservoir is also explained by the sharp fluctuations in the water horizons of the reservoir and the reservoir when the reservoir is drawn down, part of the suspended sediment is carried downstream and deposited in areas with lower velocities.

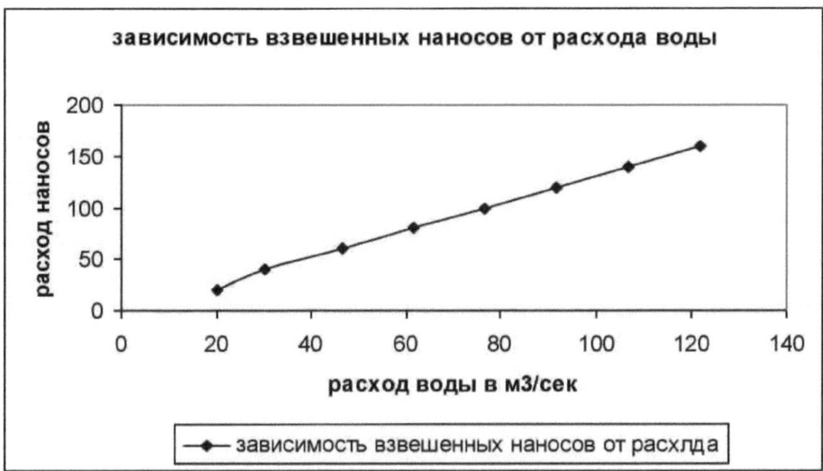

2.7. Channel deformation in the downstream channel

The release of water from the reservoir unloaded with sediment results in a general deformation of the river channel in the lower reach. Changes in the position of the average river bed bottom in the lower reach were determined by vertical survey of 25 cross sections in a 29.4 km long section from the tunnel outlet portal.

Scour and debris volume calculations for the vertical survey sections are presented in Tables 12,13, 14.

Only the section between gauges 19 and 26 at distances of 9 and 29km was found to be deformed. Table№ 12 shows that the total volume of scour was 91.8tm^3 and debris 78.1tm^3. The greatest bottom depression was observed between gauges 19 and 23 and is 0.1 - 0.15m, in other gauges on the contrary, the river channel is overburdened, which is most pronounced in gauges 19 - 22, where the average height of the overburden reaches 0.17 - 0.22m.

Further deformation of the river channel is shown in Table No. 13.

The total volume of riverbed debris was 44.7tm^3. The volume of scour is only 19.9tm^3. The largest volume of scour was observed in the section 13 - 19 and 21 - 23 gauges and the average depth of scour reaches about 0.1m, while a significant section of the river from 17 to 25 gauges was subjected to debris and the average height of the debris was also 0.1m.

During the period 2005 - 2006 significant deformation occurred in the section between gauges 13 - 24, where the average depth of channel scour between gauges 18 - 24 is 0.1m and the average height of debris in gauges 16 - 17 does not exceed 0.05m. The total volume of scour amounted to 44.3tm$^{(3)}$and 31.2tm^3, i.e. the channel sediments were washed away.

More significant deformation of the river channel has occurred in the section between stations 16 and 26, with intensive scour between stations 16 and 20 and debris between stations 24 and 26.

The average depth of scour is 0.1m, and the average height of the blockage is 0.08m, the total volume of scour is 147.34tm^3 and blockage is 141.04tm^3. These deformations of the river channel occur under conditions of different duration and size of discharge discharges from the reservoir.

Total water runoff

years	Scour volume thousand m^3	Debris volume, thousand m^3	Maximum discharge flow rates, m^3/sec	Average discharge flow rates, m^3/sec	Total water runoff per year, mln.m^3	Discharge duration, month
2004			200	60	1050,0	3
2005	96,0	78,1	65	60	970,0	4
2006	19,0	44,7	93	60	890,0	2,4
2007	44,3	31,2	95	60	960,0	3,5
2008	147,04	141,04	114	80	954	5

The table shows that maximum discharge rates cause more significant deformations of the river channel.

The nature of the riverbed deformations in the sections is shown in Figure 24 - channel deformations occur both due to lowering of the river bottom, erosion of the channel sides and debris in the river floodplain in places of large expansions.

As the deformation analysis shows, there is a general tendency for channel sediments to wash away from the upper sections over 11680m and accumulate in the lower sections over 14200m.

The fractional composition of bottom sediments in the downstream section over a number of years is summarized in Tables№ 17, 18, 19, 20 and 21. It can be established that the average diameter of the fractions from station 13 downstream to station 26 decreases from 65mm to 49mm in 2002, from 59mm to 28mm in 2006 and from 60mm to 34.4mm in 2007, which can be traced throughout the entire observation period since the beginning of reservoir operation; however, in some sites there is an increase in the average sediment diameter by years, so in site 19 from 68mm in 2002 to 83mm in 2003 to 71.5mm in 2006 and to 53.5mm in 2007; in site 21 from 21mm in 2002 to 31.5mm in 2006 and to 66mm. in 2007; in station 26 from 49mm in 2002 to 64mm in 2003 and to 34.4mm in 2007.

The increase in coarseness is due to self-paving of the channel bed. Some decrease in particle diameter in Stream 26 and elsewhere is likely due to deposition of finer fractions brought from upstream eroding sections of the river.

Observation of the progress of debris and siltation of the reservoir basin by measurements and leveling of the upstream gauges was started in October 2007. The results of these works allowed to establish average values of siltation layer thickness.

2.8. Siltation of the reservoir bowl

From the data characterizing the operating conditions of the reservoir, it can be concluded that the character of the currents in the zone of the backwater curve wedging out, depending on the filling of the reservoir, was complex. The established zones of the backwater curve wedging out, during the period of intensive movement of bottom and suspended sediment, are shown in Table 22.

Backing curve wedge-out zones

years	Filling reservoir	Support length	Backing wedging zone	
			between stations	section length, m
2004	728,0 - 744,8	5,0 - 11,5	4 - 9	6,5
2005	728,0 - 747,5	5,0 - 12,0	4 - 10	7,2
2006	744,0 - 754,2	11,2 - 15,0	9 - 12	3,8
2007	728,0 - 754,8	5,2 - 14,2	4 - 12	9,0
2008	728,9 - 756,75	5,5 - 14,8	4- 12	9,3

The section between gauging stations 4 and 11 at a length of 10km was most affected by siltation.

At that, the most intensive process of blockage and siltation of the bowl occurs in its middle part along its width (in the zone of the domestic channel before flooding) and in the initial part along the length of the reservoir, which is illustrated by the data in Table 23.

Siltation of the thicket along its length

Periods 1	Sashes 2	Average siltation thickness m 3	Width of siltation area m 4	Siltation layer thickness is related to reservoir depth 5
	4 a	0,1 -0,12	600-800	1 /230 H
2004 - 2007	5	0,1 -0,14	400-500	1 /240 H
	6	0,10	800-900	1/300H
	7	0,10	500-600	1 /250 H
	8	0,08	400-600	1 /240 H
	9	0,25	400-500	1 /60 H
	10	0,20	130-150	1 /55 H
2007 - 2008	6	0,02	800-1000	1/1500H
	7	0,014	600-700	1 /1800 H
	8	0,01	400-650	1 /1900H
	9	0,10	500-600	1 /150H
	10	0,30	150-170	1 /37 H

where H is the depth of water filling in the corresponding gauges at the NPG.

Siltation counts by station for these periods are given in Tables 23, 24, 25.

It should be noted that the transect survey was not done in sufficient detail due to more than one drawdown of the reservoir, taking into account that determination of the siltation layer by measuring depths along the transects gives significant errors.

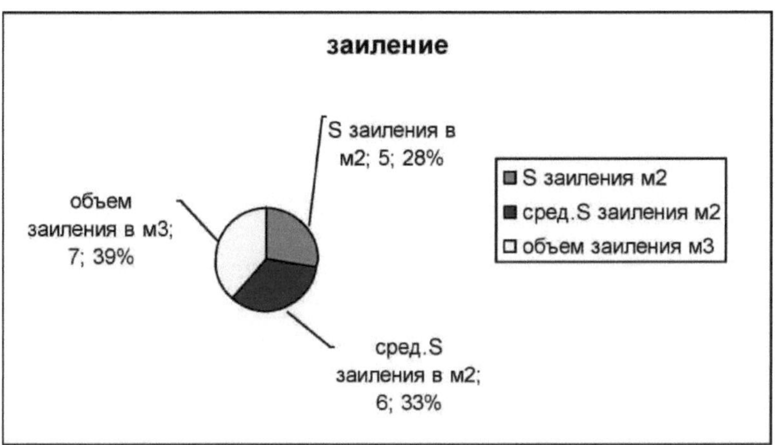

Figure 21.

General notes on the tables:

1 . during the leveling of gauges, a survey of the reservoir thicket and determination of the siltation layer was carried out;

2 determination of siltation layer thickness was performed at a distance of 100 - 300m from each other.

Determination according to Kochkorskaya G.M.P. and the volumes of suspended sediment calculated from the curve of dependence were: for the period 2005 - 420.0 for 2006 - 257, for 2007 - 103.2 and for 2008 - 23 tons.

The calculation of the runoff of sediment load is based on the formula of Y.A. Nikitin (due to the lack of observations of sediment load at the Kochkor gauging station).

The runoff of transported sediment was 137.4 tons in 2002-04, 45.4 tons in 2005, 30.6 tons in 2006, 5.0 tons in 2007 and 9.1 tons in 2008.

Table 27 shows the total volumes of suspended and transported sediment entering the Kochkor gauging station and the volumes obtained from the vertical survey data.

Total sediment load

years	total volume of siltation, t m^3	Suspended sediment runoff at the Kochkor gauging station, t m^3	runoff of transported sediment calculated	total sediment load, t m^3

			according to Nikitin, t m^3	
2004	771,0	772,0	85,5	797,5
2005				
2006	713,1	555,0	50,7	605,7
2007				
2008	220,8	88,0	5,7	93,7

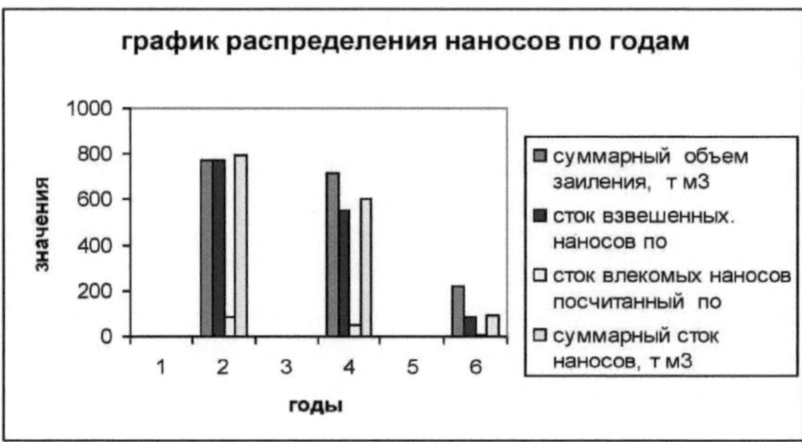

Figure 22.

Comparing the total volume of incoming sediment with the volume obtained from the survey data, it can be established that the latter is approximately 2 times greater than that calculated from the observations at the Kochkor GMP.

The average annual discharge of sediment and suspended load from the Chu River is about 244,000 m^3 or 1/1930 of the full capacity of the reservoir. It should be noted that the years under consideration were average years in terms of availability; with higher floods the runoff of sediment and suspended sediment would increase, which would also affect the change in the indicator characterizing the service life of the reservoir.

The fractional composition of sediments in the reservoir bowl, upstream of site 5, is given in Tables 27 -31.

Analysis of the data in Tables 27 - 31 shows that the average coarseness of fractions increases upwards along the reservoir. In addition, it should be noted that the composition of sediments along the length of the reservoir changes depending on the reservoir filling. Thus, if the average diameter of fractions in section 6 was equal to 25mm, and in section 11-83mm (in 2008 the maximum filling was 747.0), the average diameter decreased to 19mm in section 6 and 35.0mm in section 13, respectively.

2.9 Calculation of sediment particle size distribution

Analysis of data on the fractional composition of suspended sediment (Table 32) for 4 characteristic years shows that the bulk of suspended sediment is less than 0.01mm (34.8%).

Fractional composition in % by weight

Year of observation	Fractional diameters in mm			
	1 ÷ 0,25	0,25 ÷ 0,05	0,05 ÷ 0,01	0,01
2006	24,6	25,0	19,0	30,5
2007	24,2	26,8	14,2	34,8
2008	15,5	33,0	8,2	38,8
average	26,8	22,2	16,2	34,8

No observations of bed load runoff were made at the nearest posts to the dam site, so the runoff into Ortokoto Reservoir was assumed to be 10% of the suspended load runoff. The runoff of suspended load was calculated using the formula of Nikitin N.A., which was determined in the dimensions shown in Table 33.

Bottom sediment runoff

Years	Average annual river discharge, m^3/sec	Suspended sediment runoff, thousand tons	Runoff of transported sediment, thousand tons	Note
2004	33,4	337,0	64,2	Calculation of runoff of sediment load was calculated using the formula of Nikitin N.A.
2005	33,1	420,0	45,4	
2006	28,0	275,7	30,6	
2007	23,0	103,2	5,0	
2008	23,1	123,0	9,1	

Chapter 3: Channel processes on the Naryn River

3.1 Siltation of reservoirs on the Naryn River

The siltation of the Uchkurgan reservoir, the siltation of the Toktogul reservoir and the distribution of sediments in its water area were analyzed on the basis of observations of the solid flow of the Naryn River.

The Naryn River cuts through a ridge of Jurassic Cretaceous and Tertiary sandstones, clays and conglomerates. The high gradients and flow velocities of the river carry huge quantities of suspended sediment, large bedloads and even boulders. The main discharge of both suspended and bed load occurs during floods. In terms of mineralogical composition, the suspended load is represented by quartz, granite, feldspar, silicon, biotite, ore minerals and others. The development of the Naryn's water resources began in the late 1950s, when the Naryngidroenergostroy construction organization was established for the construction of the Uchkurgan hydroscheme and later subsequent hydroschemes in the Naryn basin. Five HPPs have been built directly on the Naryn, Table 3.1 gives the characteristics of the hydrosystems with reservoirs.

The first-born in the Naryn HPP cascade is the Uchkurgan hydrosystem commissioned in 1961, and the Toktogul hydrosystem commissioned in 1975 is the largest both in terms of capacity and reservoir volume. Chronological course of putting the reservoirs into operation corresponds to the course of their siltation. Let us start consideration of siltation processes from the Uchkurgan reservoir.

The construction of the Uchkurgan HPP dam in a narrow canyon formed a reservoir with a length of about 17 km and a mirror area of 4 km^2, a volume of 52.5 million m^3, and a usable capacity of 20.9 million m^3 designed to regulate daily runoff and create a water reserve for irrigation.

Characteristics of hydrosystems of the Nizhnenaryn cascade of HPPs

Table 3.1.

Name of HPP	Year of commissioning of the first unit	Calculation. pressure H, m	At the station. capacity MW	Reservoir			
				Mirror area, km²	Total volume, mln. ı. .3 M	Usable volume, ı. .3 M	Extended Tb KM
T octogul	1975	140	1200	284,3	19500	14000	80
Kurpsai	1981	91,5	800	12,0	370	35	40
T ashkumyrskaya	1985	53	450	7,8	140	16	20
Shamaldysayskaya	1990	26	240	2,4	40	5,7	12
Uchkurgan	1961	29	180	4,0	53,4	20,9	17

In order to preserve the useful capacity of the reservoir and to ensure the transit of sediment to the lower reservoir during the period of solid runoff concentration (April-August), Saohydroproject, with the participation of station personnel, developed measures for a rational reservoir operation regime. The measures envisaged lowering the water level in the upper embankment by 5m relative to the normally backed-up level (NBU) and flushing with concentrated flows of 1000...1500M^3 . It was supposed that under such a regime the "dead" volume of the reservoir would be filled in 5.6 years, while the volume of the useful prism would remain untouched.

During the period from 1962 to 1973, before the Naryn River was dammed at the Toktogul HPP site, the Uchkurgan reservoir received domestic river sediment runoff. Special measures on rational regime of the reservoir operation were hindered by water intake into canals (Big Fergana, Uch-Kurgan, North-Fergana and Akhun-Babaev) located downstream of Uchkurgan HPP, as a result of which the reservoir is significantly silted up.

Observations of changes in the reservoir capacity were carried out in subsequent years [8]. The works included measurements of the Naryn river channel at the reservoir site and calculation of reservoir volumes.

The change in the capacity of the Uchkurgan reservoir by years has the following dynamics: in 1957. - 54mn.m3, in 1963. - 42mn.m^3 in 1968. - 24mn.m^3, since 1978 and up to now - 15.5 mn.m^3. Thus, the full capacity of the Uchkurgan reservoir is 15.5mn m^3, which is almost 3.5 times less than its design capacity, and the usable capacity is 11.0mn m^3.

Comparison of the longitudinal profiles from the 1957.1989 surveys shows that the greatest thickness of sediment deposits up to 25m in depth are found at a distance of 1880.2735m from the dam site.

According to the results of the analysis of the particle size , the thickness of these sediments in this area is formed by sediment with a percentage of up to 62.5% of dust fractions with a diameter of less than 0.05 mm. In the upper part of the reservoir, the thickness of the sediment decreases and is formed mainly by sediment d = 0.004...0.04mm,

interspersed fractions with an average diameter of 0.1 mm with inclusion of stones d = 0.3m.

At present the reservoir is almost completely filled with sediment and this fact has a negative impact on the reservoir operation regime and the safety of the dam with regard to flood events, in particular, according to the results of the survey carried out by us in 2000, it was revealed that out of 8 bottom spillways of the dam, 3 spillway heads are silted up. The task of cleaning the Uch-Kurgan HPP reservoir on the Naryn River is urgent for its operation.

The volume of solid phase discharge from small watercourses at the Uchkurgan reservoir site is insignificant and cannot significantly affect its siltation. In addition, under proper operation the reservoir can self-clean. Therefore, in our opinion, once purification will ensure normal operation of the reservoir for many decades.

We do not consider siltation of the upstream reservoirs of the Shamaldysai, Tashkumyr and Kurpsai HPPs, as the Naryn River sediment load is already fully retained by the Toktogul reservoir at the beginning of their construction.

Consider the siltation of the Toktogul reservoir, the first in the Nizhnenaryn cascade of hydroelectric power stations with a total volume of 19.5 billion $m^{3(3)}$. The analysis of suspended sediment discharge and runoff into the Toktogul reservoir was based on observations carried out since 1964 at the Uch-Terek gauging station located upstream of the reservoir.

According to data from the Kyrgyz Hydrometeorological Center, a series of mean monthly sediment loads were compiled for the Naryn River, near Uch-Terek station. Unfortunately, some data were missing (marked in brackets) due to the lack of material resources at the Kyrgyz Hydrometeorological Center and unfavorable weather conditions (ice cover, etc.) at the time of the measurements, which did not allow this type of work to be carried out. The missing data were obtained by interpolation of available data.

The result is a series of 27 years, from 1964 to 1992, excluding 1976 and 1988, which is presented in Table 3.2. The mean annual discharge of suspended sediment R_0 and the coefficient of variation C_{vr} were calculated using known formulas.

Average monthly sediment load on the Naryn River, near Uch-Terek station, kg/sec

Table 3.2.

Year	Months												Average annual
	I	II	III	IV	V	VI	VII	VIII	IX	X	XI	XII	
1964	(10)	(10)	(70)	(60)	40	2700	440	68	13	1,4	(2)	(2)	(285)
1965	(10)	(10)	24	90	90	120	1100	580	98	37	48	(18)	(185)
1966	52	(25)	(40)	(750)	1500	7000	1900	1700	250	50	29	32	(AI)
1967	(15)	20	53	970	(2500)	2000	1200	460	85	30	9,5	17	(613)
1968	(7,2)	18	74	470	970	2500	1500	980	120	14	22	9,4	(557)
1969	18	27	510	1500	5200	5200	4400	1200	530	89	30	18	1600
1970	6,9	31	35	370	1300	1300	1900	700	210	23	15	7,4	490
1971	6,1	31	210	270	360	2500	880	470	65	28	31	16	410
1972	10	10	95	150	590	1600	840	920	71	33	50	И	360
1973	(Ю)	(20)	42	600	1400	3200	2000	670	210	3,2	(16)	(10)	680
1974	(15)	(20)	5,9	140	370	630	680	330	43	7,7	(Ю)	(20)	190
1975	(8)	(Ю)	6,4	220	220	1900	590	570	64	(Ю)	(15)	(4)	(301)
1977	(2)	(Ю)	(75)	150	(900)	(970)	600	400	62	71	46	8,3	(275)
1978	3,5	1,8	69	1100	1400	930	1000	1000	35	9,1	8	20	460
1979	4,1	6,9	110	1200	1000	2700	(1700)	880	85	170	23	(18)	(658)
1980	(2)	(17)	(75)	180	1700	930	640	380	46	14	(6)	(8)	(333)
1981	15	35	49	230	1300	1100	1100	270	41	19	21	6,4	350
1982	(10)	(6)	35	340	340	130	360	450	38	12	10	(2)	(144)
1983	1,1	7,9	37	160	880	1300	1100	1100	170	49	14	10	400
1984	5,8	12	110	520	320	720	480	1100	46	15	14	2,8	280
1985	(10)	(10)	41	430	280	780	520	230	(10)	15	17	(10)	(196)
1986	2,1	16	32	200	280	940	880	360	54	28	10	7	230
1987	6,7	8,2	120	240	1200	3000	2900	1100	110	100	34	21	740
1989	(10)	(10)	(НО)	(610)	(2400)	1800	1400	450	90	26	23	12	(578)
1990	45	35	280	720	2100	1400	1200	520	270	43	24	24	560
1991	(15)	(15)	200	320	440	2300	1400	670	170	24	25	(10)	(466)
1992	(12)	(10)	130	480	580	920	1300	310	63	22	15	15	(321)

As a result of calculations based on research, we obtained a mean annual sediment load R0= 473kg/s and coefficient of variation C_{vr} = 0.655.

The results of the FLIKE program (Log Pearson III) are shown in Table 3.3.

Sediment load at various levels of supply

Table 3.3.

Provisioning *P%*	0,1	0,5	1,0	2,0	10,0	50,0
Sediment load R_o, kg/s	3548	2335	1920	1560	894	392

The fractional composition of the sediment is taken from the gauging stations nearest the Uch-Terek gauging station. Table 3.4 shows the fractional composition of the Naryn River sediment at the Nizhnenaryn section of the HPP cascade.

Average particle size distribution of suspended sediment in the Naryn River

Table 3.4.

The post	Years of observation	Number of samples	Sediment coarseness in mm and content as % of total load			
			> 0,20	0,20 - 0,05	0,05 - 0,01	<0,01
p. Kekirim	1938-1943, 1956-1960	202	2,9	21,2	17,7	58,2
c. Alekseevka	1933-1935, 1937-1942 1948, 1950-1952, 1954, 1957, 1958	74	1,8	21,9	21,5	54,8
st. Verkhne-Uchukrganskaya	1953-1957		2,8	25,0	19,1	53,2
kish 1. Uchkurgan	1929, 1931-1944, 1946-1962	124	2,4	15,2	27,2	55,2

Table 3.4 shows that the average particle size distribution of suspended sediment in the river Naryn in this section of the river is almost unchanged, so for further calculations we have adopted the hydrological distribution of the village of Alekseevka. Alekseevka.

The total approximate estimation of the duration of the period of reservoir filling with sediment is analytically derived and is not given here.

The assessment of the distribution of sediment over the reservoir area is based on the reservoir flow velocity v_{cp} and the deposition velocity of particles of a given size V_B suspended in the flow. The relationship between v_{cp} and v_B.

Knowing the rate of deposition of suspended sediment and the average flow velocity in a section of the reservoir, it is possible to calculate the path *L* over which sediment of the size in question will be deposited

When calculating the siltation of the Toktogul reservoir, a number of observations of solid runoff at the Uch-Terek gauging station (see Table 3.3), located upstream of the Toktogul reservoir, were taken as a basis.

Further, assuming that the dead volume of the Toktogul reservoir is 5.5 billion m^3, and assuming that the transit portion of fine sediment discharged from the reservoir during floods is a fraction of the total sediment volume d = 5%, the possible time of siltation of the dead volume of the Toktogul reservoir was calculated.

As follows from the above, the assessment of the distribution of sediment in the reservoir area is based on consideration of the flow velocity Vcp in the reservoir and the deposition velocity of particles of a given size VB suspended in the flow.

The hydraulic coarseness of the sediment was determined from Table 9.3 of SNiP and the limiting velocity of particle fallout was determined from SNiP formulas.

In calculating the distribution of sediment over the reservoir area, 6 stations at equal distances from each other are assumed. Thus, having data on the solid discharge, coarseness, or particle size distribution of the sediment entering the reservoir, the flow rate and cross-sectional area at different sections of the reservoir, it was also determined at what distance from the backwater wedging zone the deposition of particles of different sizes would occur. The calculations are summarized in Table 3.5.

Fig.7. Sediments in the upper part of the Toktogul reservoir during the low-water period

Calculation of sediment distribution in the water area of the reservoir

Table 3.5.

Calculation parameters	Particle diameter, MM	Numbers of settlement stations (distance, km)					
		1 (0,0)	2 (2,0)	3 (4,0)	4 (6,0)	5 (8,0)	6 (IO,0)
Cross-sectional area of protected waters, m^2		1000	5000	10000	25000	50000	100000
Protected water width, m		200	500	700	1120	1570	2200
Protected water depth, m		5,0	10,0	14,3	22,3	31,8	45,5
Cf. water velocity *Vcp,* m/s		1,558	0,3116	0,1558	0,06232	0,03116	0,01558
Limit particle fallout velocity, Vk, m/s	0,2	0,283	0,318	0,337	0,363	0,385	0,409
	0,05	0,178	0,200	0,212	0,229	0,243	0,258
	0,01	0,104	0,117	0,124	0,134	0,142	0,151
Particle fallout velocity,Vb, m/s	0,2	deny.	0,004	0,113	0,174	0,193	0,202
	0,05	deny.	deny.	0,001	0,001	0,002	0,002
	0,01	deny.	deny.	deny.	0,00004	0,00006	0,00007
Length of particle fallout area,L, m	0,2	whizzes by	785,6	19,7	8,0	5,1	3,5
	0,05	whizzes by	whizzes by	4180,0	956,0	569,3	376,8
	0,01	whizzes by	whizzes by	whizzes by	32557,2	15894,1	9873,8

Based on the results in Table 3.5, the distance at which suspended sediment particles of a given fractional composition will fall out has been determined: fractions with a diameter of d = 0.2mm will fall out at a distance of approximately 2.8km, d = 0.05mm - 8.2km and d = 0.01mm - 19.9km.

The following conclusions can be drawn from the results of the data obtained:

On the Uchkurgan Reservoir:

- Comparison of the results of cross-sectional measurements with turbidity studies shows that the volume of siltation is almost entirely determined by the volume of suspended sediment; bottom sediment does not have a significant effect on the siltation of the reservoir.
- On the basis of study of available materials on reservoir siltation and its cleaning and our surveys (2000) it was established that reservoir cleaning remains an urgent task. In order to carry out

cleaning measures it is necessary: to conduct bathygraphic survey of the reservoir; to develop a cleaning project.

On the Toktogul Reservoir:

- the period of siltation of the dead volume is rather long, more than 350 years, but siltation of the reservoir mouth may lead to bottom rise.
- The deposition of even the finest sediment fractions will occur at distances of up to 20 km from the mouth, i.e. within the Ketmen-Tyubinsk valley.
- In the foreseeable future, sediment load will not reach the dam site and cannot affect the operation and safety of the Toktogul HPP.

3.2 Parameters of solid flow on the Naryn River at the site of Uchterek village. Uchterek between Kambarata HPP-1 and HPP-2

The formation of the Naryn River load is mainly due to the washing away of fine sediment from the river basins and, to a much lesser extent, to channel scour, the intensity of the latter depending on the stability of the channel. Among the products of basin washout, the main role is played by sediment carried by glacial waters and sediment washed away by snowmelt and rainwater.

The waters of the rivers within the Naryn system have high turbidity. The annual course of turbidity is characterized by increased turbidity during the flood period, and there is a tendency for its two maximums - in May, during the period of intensive melting of seasonal snows, and in July-August, during the period of glacier melting. The lowest turbidity is observed during the winter months.

The proportion of suspended sediment due to glacial flushing reaches its highest values (70-80%) in the Naryn River basin, which is classified as a mixed, predominantly snow-fed river with average catchment heights of about 2500-3000m. The share of glacial flushing decreases with decreasing catchment elevation, while the share of melt and rain flushing increases.

The intra-annual distribution of suspended sediment discharge and turbidity basically corresponds to the intra-annual distribution of liquid discharge, but is even more uneven. Fluctuations in the discharge of suspended sediment are not always synchronous with fluctuations in the discharge of water. 3.1 shows that the year with the highest water content is not always the year with the highest suspended sediment load, nor will the lowest years in terms of water discharge always be the lowest in terms of sediment load.

During the autumn-winter period (October-February), the turbidity of the Naryn River fluctuates only slightly, suspended sediment loads are lowest and sediment load is only 0 to 5% of the annual discharge.

In the spring period, from March to June, turbidity increases sharply, with turbidity values reaching higher values on the rise of the flood than on the fall. The time of passage of the highest average monthly discharge of suspended sediment is not uniform along the length of the Naryn, as it depends mainly on the height of the catchments, and usually coincides with the period of maximum discharge.

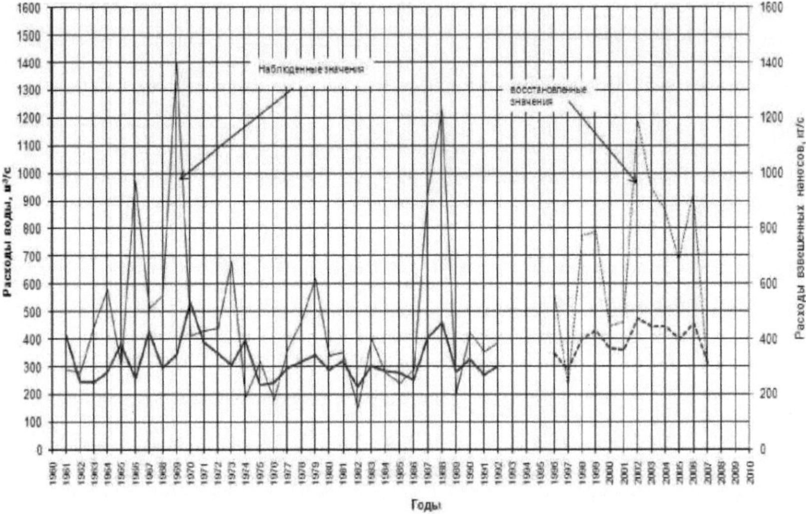

Figure 8. Graph of the relationship between average monthly expenditures

A plot of the relationship between mean monthly discharge of suspended sediment and water discharge $R_{B(3)B}$ = f(Q)was constructed for the years of observation indicated (Fig. 3.2).

The multiyear series of suspended sediment loads was obtained by reconstructing them from this graph and the average monthly water discharge for the period from 1993 to 2007.

The average monthly and average annual discharge of suspended sediment from the Naryn River at the Uchterek gauging station is given in Table 3.5. The average annual discharge of suspended sediment for 1961-2007 is 480 kg/s, which amounts to 15.2 million tons or 12.6 million m^3.

The highest average annual discharge of suspended sediment in the Naryn River in the series of observations was recorded in 1969 at 1400 kg/s with an annual flow of 44.2 million tons, and the lowest in 1982 at 150 kg/s with a flow of 4.7 million tons.

The main runoff of suspended sediment occurs during the flood period - April and August - and averages 94% of the annual runoff.

The average annual discharge of suspended sediment during this period is 1000 kg/s. The highest observed mean monthly discharge in the long-term average is in June and equals 1700 kg/s.

In some years, the maximum suspended sediment discharge occurs in May (5 times), July (2 times) or August (2 times).

The highest average monthly discharge of suspended sediment for 1961-1987 and 1989-1992 was observed in June 1969 and amounted to 5200 kg/s, while the lowest for the flood period - 35 kg/s - was observed in April 1962.

Taking into account the low inflow of the Naryn River section between the sites of the hydrological post at Uchterek village and Kambarata HPPs No. 1 and No. 2, data from observations at the Uchterek gauging station were taken to characterize the suspended sediment flow of the Naryn River to the proposed Kambarata HPPs. Measurements of flow rates and fractional composition of suspended sediment, and daily water samples for turbidity were taken at the site of the village of Uchterek from 1961 to 1961. Uchterek from 1961 to 1987 and (with omissions) from 1989 to 1992. No observations of turbidity were made after 1992.

The most clarified water in the Naryn River at the gauging station near Uchterek village is observed from November to February. The sediment load during this period is 1.6 % of the annual load.

Water turbidity of the Naryn River at the gauging station near Uchterek village averaged 1.4 kg/m^3 for the period from 1961-1987, 1989-1992.

The intra-annual distribution of turbidity corresponds to the intra-annual distribution of suspended sediment runoff (Table 3.6).

Water turbidity in the period from April to August averages 1.9 kg/m^3. The highest turbidity is mainly observed in June and averages 2.2 kg/m^3 in this month. The highest average monthly turbidity for the period 1961-1987, 1989-1992 was observed in May 1969 and was equal to 5.6 kg/m^3.

The fractional composition of suspended sediment as measured at the Uchterek gauging station in 1964-1969, 1974-1978 and 1983-1986 is shown in Fig. 9. Uchterek in 1964-1969, 1974-1978 and 1983-1986 is shown in Fig. 9.

The main mass (96%) of suspended sediment is made up of particles with a diameter of less than 0.25 mm. At the stage of justifying materials, the flow of suspended sediment from the Naryn River to the Kambarata hydropower stations was based on the percentage ratio of suspended and suspended sediment obtained from measurements in 1962-1966 on the Naryn River at the Karasuisky gauging station. This ratio amounted to 7 %.

In 1985 and 1986, Sredazgidroproekt carried out observations of the flow of sediment load from the Naryn River at sites near Uchterek village and upstream of the mouth of the Kekemeren and Kekemeren rivers. The Sredazgidroproject monitored the flow of sediment from the Naryn River at the sites near the village of Uchterek, upstream of the mouth of the Kekemeren River and the Kekemeren River at the mouth.

Measurements of sediment discharge were carried out during the flood period from May to September. The sediment was sampled using a Hydroproject net trap with an inlet opening of 15x15 cm.

In the Naryn River at Uchterek village the movement of drift sediment begins at a discharge of $200m^3/s$ and the sediment flow band varies between 50-65m. At the site upstream of the mouth of the Kekemeren River, the movement of drifting sediment is noted at water flows of more than $100m^{(3)/s}$, with the sediment flow band varying from 15 to 50m.

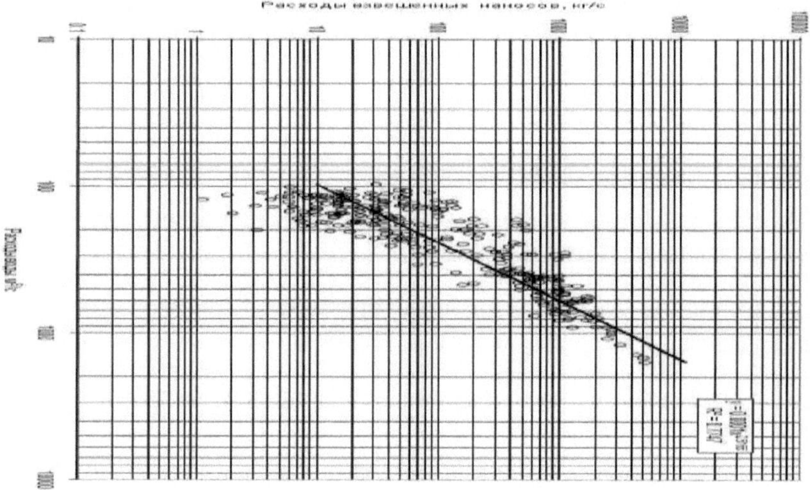

Figure 9

The results of measurements of sediment load of the Naryn and Kekemeren rivers for 1985 and 1986 are presented in Table 3.6.

Information on the solid flow of the Naryn River.

Table 3.6.

The river is an observation stem	1985 г.		1986 г.	
	Runoff of transported sediment thousand tons	% of suspended load runoff	Runoff of transported sediment, thousand tons	% of suspended load runoff

p. Naryn - above the mouth of the Kekemeren River	310	5,7	300	4,0
p. Kekemeren river - mouth	85	34	69	8,0
p. Naryn River - below the mouth of the Kekemeren River	390	6,9	370	4,3
p. Naryn - village. Uchterek	600	8,0	-	-

The flow of sediment load from the Naryn River at the site of Uchterek village in 1986 could not be calculated due to the lack of dependence of measured sediment load on water levels and suspended sediment loads. It was not possible to calculate the flow of disturbed sediment in the Naryn River at the Uchterek village site in 1986 because there was no dependence of measured discharge on water levels, discharge and suspended load. On average for 1986, the measured discharge of drift sediment at this site is 3.6 % of the discharge of suspended sediment.

Thus, the percentage ratio (7%) of runoff (7%) of the Naryn River to the Kambarata HPP-1 and 2 sites adopted at the stage of substantiating materials is confirmed by observation data from 1985 and 1986.

The runoff of the Naryn River to the sites of Kambarata HPPs 1 and 2 is assumed to be 1.06 million tons or 0.53 million m^3 per year.

Observations in 1985-1987, 1988, 1989, 1991 and 1992 on the fractional composition of the Naryn River sediment at the hydrological station near the village of Uchterek also confirmed the curve given in the Technical Project and obtained on the basis of observations in 1962-1966 at the Karasuisky gauging station (Fig. 10).

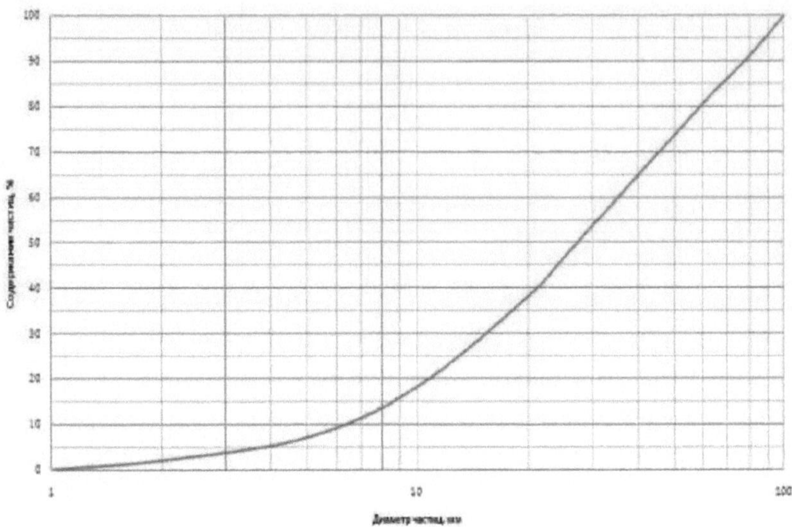

3.3 Methodology of siltation calculations

In construction design practice, the main method of calculating reservoir siltation is the balance method, which takes into account the mutual influence of the flow and its bed. The method is based on the joint solution of the equations of fluid motion and the balance of solid material, which makes it possible to obtain not only the total amount of sediment in the reservoir, but also the distribution of sediment over the reservoir area within the channel and floodplain parts. Preparation for the siltation calculation includes the following operations.

The design year (or series of years) is divided into several time intervals for which the reservoir water level and incoming water discharge and turbidity can be assumed constant.

The reservoir is divided on a large-scale plan into design sections, the number of which depends on the availability of baseline information. This takes into account the outline of the reservoir in plan view and the increase in length of the backwater section as a result of siltation. The increase in water level in the zone of the initial upwelling due to sediment deposition can be determined by the ratio:

$$\Delta Z = L \cdot i_{np} \quad (3.1)$$

where L is the initial length of the buttress, m.

The length of the backstop at the limiting slope iП p. is determined by the formula:

$$L_{np} = L + \frac{\Delta Z}{i_{быт.} + i_{np}} \quad (3.2)$$

The ultimate slope of the water surface is calculated by the formula:

$$i_{np} = \left(\frac{\varepsilon \cdot \mu}{24} \right)^{0.85} + \frac{n^2}{i^{0.5}} \quad (3.3)$$

where ε is the average turbidity of channel-forming sediment fractions during the flood period of the design year, g/m^3;

n - roughness coefficient.

The length L at which the backwater from the dam acts after the siltation process is complete is typically 2:2_5 lengths of the reservoir prior to the onset of siltation.

The width of the channel formed in sediments can be taken equal to the natural width of the river. The length of design sections can be equal to 2÷5 average widths of the reservoir. Places of sharp narrowing or widening of the valley should also be distinguished as design sections.

Cross-sectional profiles are constructed at the boundaries of the design sections. In reservoirs, where islands, floodplain and terraces fall into the zone of permanent and periodic flooding, schematization of cross-section should be made with allocation of channel and floodplain compartments. In the dammed part of the reservoir with a depth of more than 15m, the floodplain compartment is not expedient.

Within each bay, the widths and mean bottom elevations are determined.

$$h = Z - H_{cp} = Z - \omega / B \quad (3.4)$$

where: h - average bottom mark, m,

Z - water level mark at which the live section area ω was determined, m,

B - width of the compartment, m,

Ncr - average depth, m.

Schematization of the channel cross-sectional profile by rectangular compartments can lead to distortion of hydraulic resistances and, consequently, of the calculated water levels. Correspondence of calculated levels to natural levels at the same water discharge, which is especially important for the backwater wedging zone, is achieved by selection of the calculated roughness coefficient n.

The calculated value of the coefficient n is selected by calculating the free surface curves (without backwater) according to the hydraulic characteristics of schematized cross-sectional profiles in comparison with the natural . The natural position of the free surface curve is established by the dependences Q = f(Z), which are constructed in a number of gauging stations according to the results of hydrometry or hydraulic calculations. Calculation of natural free surface curves and also backwater curves is performed according to the Shezy-Manning formula.

The transport capacity or ultimate saturation with channel-forming sediment ε (g/m^3) at each design section is determined from the hydraulic conditions at the outlet gauging station using the formula:

$$\varepsilon = \frac{24 \cdot V^3_{cp}}{\mu \cdot H_{cp}} \qquad (3.5)$$

Where

The calculated characteristics of the mechanical (particle size distribution) composition of suspended sediment are determined for 3 fractions:

the first - with particle diameter larger than 0.05mm, the second - with particle diameter 0.05÷0.01MM,

third - with particle diameter smaller than 0.01mm.

The coefficient α is calculated for the first two fractions, while μ is defined for the average particle diameter of this fraction. Particles of the third fraction (finer than 0.01mm) are assumed to settle uniformly outside the sediment bar over the entire reservoir area until the average flow velocities in the reservoir reach 0.20m/s or more. From this point, fine particles can be transported downstream.

Coordinates of the curve of limiting saturation of the flow with particles> 0,05 mm are calculated by the formula:

$$\varepsilon_{np} = \alpha \cdot \frac{V^3_{cp}}{H_{cp}} \qquad (3.6)$$

The equation for the limiting saturation of the flow with 0.05÷0.01MM particles (sediment fraction II) is written in the same way.

The calculation of suspended sediment load in the reservoir consists of two successive stages, which are repeated until the transport of all suspended sediment to the downstream reservoir is restored:

- determination of the free surface curve for average or characteristic

water flow rate of a given design time interval,

- calculation of bottom deformations at the outlet stations during the same design time interval.

Curves of the free surface of the reservoir (from the hydroscheme site to the site where the design level at a given discharge coincides with the natural level) are calculated at a given water level at the dam and discharge at each design site.

At the same time, the distribution of water discharge between channel, floodplain and other compartments identified in the cross-sections of the reservoir is determined. It can be taken as proportional to the capacity of each compartment:

$$\frac{B \cdot H^{5/3}}{n} \quad (3.7)$$

The calculation of bed deformations (sediment deposition or scouring) is carried out starting from the upper section - from the point where the backwater is wedged out - downstream.

Turbidity of water entering the upper section (in the section where there is no backwater) is taken in accordance with its domestic value, average for the calculated time interval. The turbidity at the outlet of each section is determined according to the dependence ε = f (V, H) for each of the fractions into which the sediment particle size distribution curve is divided. Here V is the flow velocity, m; H is the flow depth, m.

The height of the sediment layer Δh at each design section is determined from the sediment balance equation, which also takes into account the transport capacity of the flow:

$$\frac{q \times \varepsilon}{\gamma \times l} t = \int_0^h \frac{dh}{1 - \frac{\alpha}{\varepsilon} \cdot \frac{q^3}{(H_0 - h)^4}} \quad (3.8)$$

where: q - specific water flow rate in a compartment with width b: q = Q/b, m^3/s,

ε - turbidity of water at the site inlet, g/m^3,

γ - sediment density, g/m^3,

l - length of the design section, m,

t - calculated time interval, sec,

α is a constant related to the hydraulic size of each sediment fraction,

But - depth at the lower boundary of the site before the onset of deformations, m.

Integration of equation (3.8) gives here:

$$A = \eta + \frac{1}{4\beta}\left[\ln\frac{(\beta-1)[\beta(1-\eta)+1]}{(\beta+1)[\beta(1-\eta)-1]} - 2arctg\frac{\eta\beta}{1+\beta^2(1-\eta)}\right] \quad (3.9)$$

$$A = \frac{\varepsilon_0 q t}{\gamma H_0}; \quad \eta = \frac{\Delta h}{H_0}; \quad \beta = \frac{\varepsilon_0}{\alpha q^3} H_0$$

where: ε0 - water turbidity in the inlet station, g/m^3,

Δh is the change in depth due to sediment deposition, m.

The value of Δh determined at the outlet gauge applies to the entire length of the design section.

Turbidity at the site outlet is determined by the formula:

$$\varepsilon_{вых} = \varepsilon_0 - \frac{\gamma \cdot l \cdot \Delta h}{q \cdot t} \quad (3.10)$$

The conveying capacity of the flow at the outlet of the design section is determined from the saturation limit curves.

The condition for sediment deposition is:

- for the first fraction εprd.I < inh.T;
- for the second fraction εp(pe)(e).II < εβx.II.

If εpred.I = εin.T, all sediment (both I and II fractions) is transported to the next calculation section. If εpred.P> εin.∏ or εpred.I= εin.I only the second fraction is transported to the next section. The inequality Cpree.1> ε(β)(x).I is a condition for the flushing of sediment deposited in the preceding calculation time intervals.

In all these cases, the numerical value of "α" is taken :

When only the first fraction is deposited (vpred.T< εβx.I) - α = αI.

The same in the case of flushing of settled sediment (in prep.1> εβx.I) - α = αI.

When both fractions are deposited, α = αI + II. to determine αI + II from the observed data, a relationship is constructed:

$$\varepsilon_{I+II} = f\left(\frac{V}{H^{0..33}}\right)$$

and the equation of the upper envelope of the point field, which has the form:

$$\varepsilon_{I+II} = \alpha_{I+II} \times \left(\frac{V}{H^{0..33}}\right) \quad (3.11)$$

The value of αI + II and is taken as the design value for the deposition of both fractions.

Once sediment deposition has been calculated for each time interval, the mean bottom elevations at the calculated reservoir sites are modified; the modified bottom elevations are used to determine the backwater curve at Qpec4. the next time interval, then the bottom deformation Δh due to sediment deposition for that interval is calculated, and so on until all suspended sediment transport to the downstream reservoir has been restored.

Chapter 4. Water outlet-stabilizer of water discharge from canals with turbulent flow regime

4.1 Hydrosystems with devices to prevent sediment accumulation from turbulent channels

The invention relates to hydraulic engineering and can be used for stabilization of water supply from canals with turbulent flow regime.

A water flow energy damper is known (patent RU№ 2489545, C1, cl. E02B 8/06, 10.08.2013), which includes a supply channel with a cantilever installed in the recess of a water well, and a water wall made curved in the vertical plane. The water-retaining wall is made in the form of a gate, which has in cross-section the shape of a sector, attached to the rear wall of the well by means of a hinge with a horizontal axis of rotation. The well with a niche contains a bottom gallery with a washing hole. The cross-section of the recess corresponds to the cross-section of the gate in the non-operating position (bottom).

The disadvantage of this invention is the complexity in the manufacture of this technical solution, consisting in a large number of components, as well as available moving parts, for example, in the form of cylindrical counterbalance floats, rotating gate, which will not provide durability of this system.

The closest in technical essence from the previously known is a water outlet-stabilizer of water flow from canals with a turbulent flow regime, (patent RU№ 2484203, cl. E02B /13/00, 10.06.2013), containing a water-receiving gallery, made with a sloping bottom between the supply and transit fast-flowing channels with a wave structure of the flow. The outlet along the length from the gallery is connected with a discharge pipe (channel), in the inlet section of which a gate with a lifting mechanism is installed.

The disadvantage of the known solution is low reliability, as the profile and elastic material of curvilinear plates attached by the end from below to the vibrating grid, allowing to move (bend) in the direction of gaps in the grid, can be blocked by large pumps and, in this case, will not be able to perform the designated function. That is, when the water outlet is operating, the flow of water through the gaps in the grate will not bend the plates to the required value necessary to pass the required water flow rate in the initial part of the gallery, thus will not increase the structural gap between the bars of the vibrating grate to increase the capacity of the device.

The objective of the invention is to increase the efficiency by simplifying the design and manufacturability of manufacturing of the water outlet, whereby stabilized water intake from the fast flow channel is provided by changing the hydraulic resistance of the flow.

The task is achieved by the fact that water outlet-stabilizer of water flow from canals with turbulent flow regime, including supply and transit canals, bottom water-receiving gallery, which has a water-receiving hole in the upper part, covered from above with a grate, regulating gate, in this case in the bottom water-receiving gallery is made a well consisting of two sections, equipped with regulating gates of water outlets of outlet canals, water cutoff is fixed above the grate, the end and side parts of which are blind, the grate is made with plates inclined backward to the inlet water flow, with the formation of slits located perpendicular to the reverse flow of water in the cutoff, for free flow of the inlet water flow into the well, an inclined grate is fixed in front of the inlet part of the cutoff, a vertical plate and a shield are installed in the lower part of the well.

The water flow, divided into transit and intake parts, passing at high speed along the canal bottom in the area of water intake is divided into a transit part passing over the steel water separator, and the other, intake part of water is supposed to be taken into the outlet canals. The inlet portion of the steel cutoff is designed with a smaller cross-sectional area to ensure that most of the entrained sediment leaves with the transit flow. A sloping grating is installed upstream of the steel cutoff to prevent coarse sediment and fins from entering the steel cutoff. Inside the steel water separator, the end and side parts of which are blind, the water is braked and backflow occurs. At the same time, the reverse water flow is prevented by a grid with plates inclined backwardly to the water flow, forming slits located perpendicularly to the reverse water flow in the steel water separator, due to which there is a free flow of the withdrawn water into the well. When the reinforced concrete well is filled to the level of the vertical steel plate, the water overflows through the vertical steel plate and is supplied to the consumers, the flow rate of which is regulated by gates. The vertical steel plate also serves to calm the turbulent water regime in the well and to accumulate fine sediment in the well compartment. A plate is installed at the bottom of the well, the opening of which hydraulically clears the well of sediment. This is achieved by the fact that the steel water separator with an inclined grating and a grating with plates sloping backwards to the water flow, forming slits by means of a hinge fixed to the canal wall is raised, which allows hydraulic cleaning of the well compartment by adjusting the flap and the outlet-stabilizer of the water flow from the canal with a turbulent flow regime is ready for operation again.

Based on the above, the authors consider it possible to state that the proposed technical solution meets the criterion of "Significant Differences".

Fig. 1 shows the plan of the water outlet from the fast flow channel; Fig. 2 - section A-A; Fig. 3 - section B-B.

Water outlet-stabilizer of water flow from canals with turbulent flow regime contains water flow 1, divided into transit 2 and intake 3 parts, passing at high speed in the channel 4, connected to the well

5, equipped with gates 6 water outlet of outlet channels 7, at the level of the bottom 8 of the channel 4 is installed grating 9 with inclined back to the water flow plates 10, forming slots 11, above it there is a steel water cutoff 12 with an inclined grid 13, fixed by a hinge 14, at the bottom 15 of well 5 there is a vertical steel plate 16 for calming the turbulent mode of water and a shield 17 for hydraulic cleaning of compartment 18 of well 5 and compartment 19 for water intake from well 5 to consumers.

Water outlet-stabilizer of water flow from channels with turbulent flow regime works as follows.

Water flow 1, passing at high velocity along the bottom of channel 4 in the area of water intake is divided into a transit part 2, passing over the steel water cutoff 12, and the other water intake part 3 is supposed to be taken into the outlet channels 7. The inlet portion of the steel water cutoff 12 is made with a smaller cross-sectional area so that a larger portion of the entrained sediment leaves with the transit flow 2. A sloping grating 13 is installed upstream of the steel cutoff 12 to prevent coarse sediment and fins from entering the steel cutoff 12. Inside the steel water separator 12, the end and side parts of which are blind, the water is braked and backflow occurs. At the same time, the reverse flow of water is prevented by a grid 9 with plates 10 inclined backwardly to the water flow, forming slits 11 located perpendicularly to the reverse flow of water in the steel water separator 12, due to which there is a free flow of water 3 into the well 5. When filling

of the reinforced concrete well 5 up to the level of the vertical steel plate 16, the water flow overflows through the vertical steel plate 16 and is supplied to consumers, the flow rate of which is regulated by gates 6. The vertical steel plate 16 also serves to calm down the turbulent water regime in the well 5 and accumulate fine sediment in the compartment 18 of the well 5. At the bottom 15 of the well 5 a flap 17 is installed, the opening of which hydraulically cleans the well 5 of sediment. This is achieved by the fact that the steel water separator 12 with an inclined grating 13 and grating 9 with plates 10 sloping backward to the water flow, forming slots 11, by means of a hinge 14 fixed on the wall of the channel 4 is raised, which allows hydraulic cleaning of the compartment 18 of the well 5 by means of regulation of the flap 17 and the outlet-stabilizer of the water flow from the channel 4 with a turbulent flow regime is ready for operation again. Practical operability of the proposed energy damper is obvious and it fits into the technology of irrigation equipment design.

Thus, the economic efficiency of the proposed water outlet-stabilizer flow rate is in the simplicity of the device, as well as in the combination in one technological cycle of the tasks of optimal intake of stabilized water flow rate and effective water treatment from sediments and debris.

Formula of the invention

Water outlet-stabilizer of water flow from canals with turbulent flow regime, including supply and transit canals, bottom water intake gallery, which has a water intake opening in the upper part, covered from above with a grate, regulating gate, differing in that the bottom water intake gallery has a well consisting of two sections equipped with regulating gates of water outlets of outlet canals, a water cutoff is fixed above the grate, end and side parts of which are blind, the grate is made with plates inclined backward to the flow of the inlet water stream, with the formation of slits located perpendicular to the reverse flow of water in the cutoff, for free flow of the inlet water stream into the well, in front of the inlet part of the cutoff is fixed inclined grate, in the lower part of the well is installed vertical plate and shield.

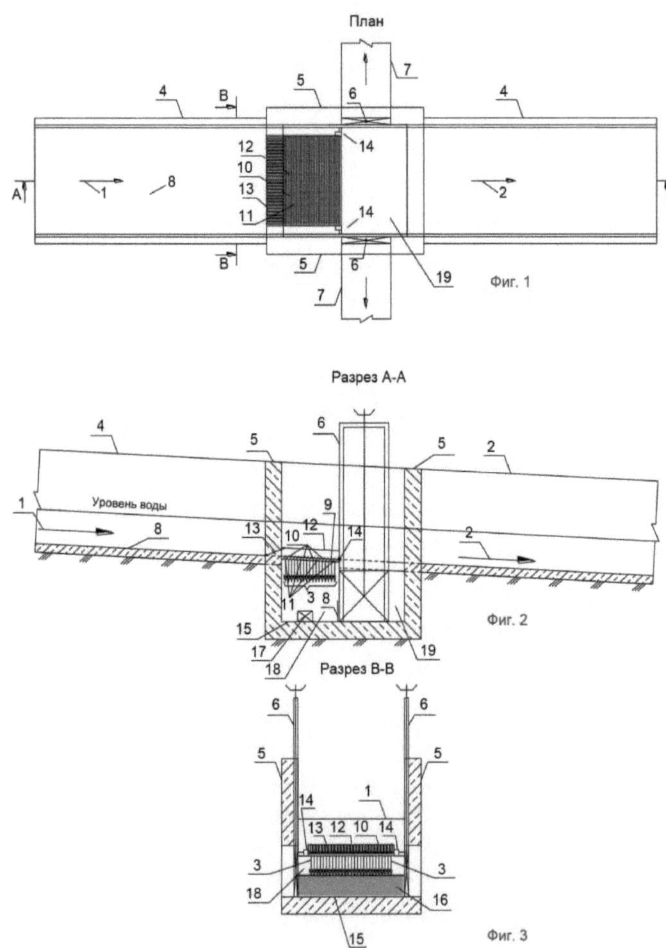

Notation:

1. Water flow
2. Transit part of water
3. Water withdrawal part
4. Channel (having a high speed)
5. Well

6. Shutter
7. Water outlet of drainage channels
8. Channel bottom (channel bottom level)
9. Grille (installed on channel 4)
10. Plates
11. slits
12. water cutoff
13. Inclined grid
14. Hinge
15. Bottom of the well (bottom of well 5)
16. Vertical steel plate (for calming the turbulent mode)
17. Shield
18. Hydraulic cleaning of the compartment (compartment 5)
19. water intake compartment (from well 5)

КЫРГЫЗСКАЯ РЕСПУБЛИКА

КЫРГЫЗПАТЕНТ

ПАТЕНТ

№ *1741*

Название изобретения: *Водовыпуск-стабилизатор расхода воды из каналов с бурным режимом течения*

Патентовладелец, страна: *Акматов А.К. (KG)*

Автор (авторы): *Сеитбаев О.О., Муканов Т.А., Каниев М.Д., Жалболдуев П.Б. (KG)*

Заявка № *20140111.1*

Приоритет изобретения *11 сентября 2014 года*

Зарегистрировано в Государственном реестре изобретений Кыргызской Республики *30 апреля 2015 года*

ПАТЕНТ ПОД ОТВЕТСТВЕННОСТЬ ЗАЯВИТЕЛЯ (ВЛАДЕЛЬЦА) на данное изобретение удостоверяет исключительное право патентовладельца на владение, использование, а также запрещение использования другими лицами на территории Кыргызской Республики.

Conclusion

Key findings

1. The control of the condition and operation of hydraulic structures makes it possible to identify the main operational difficulties, the degree and nature of which depend mainly on the head and design features of hydraulic structures, as well as on the geographical location of the hydroscheme.

2. The world has accumulated experience with existing hydroschemes and has developed sediment treatment facilities. The absence or extreme limitation of water discharges for flushing the sediments accumulated in the reservoir embankments in the fall and spring (before floods) complicate its operation, which is often impossible without mechanical removal from the reservoir.

3. For rational use of existing hydraulic structures and reservoirs created by them it is expedient to expand the study and generalization of experience of operation of hydraulic structures and hydraulic engineering structures. At the same time, the data of all types of field observations and studies should be taken into account more deeply than at present during the development of new energy and complex-energy systems. systems. It is also necessary to deepen

elaboration of operational issues when drafting integrated hydrosystems and their structures and sediment control measures.

4. At this stage all types of observations are not carried out in the Republic, partially carried out observations do not give a full characteristic of the structure operation, which does not allow analyzing processes in the reservoir.

List of references used

1. Surface water resources of the USSR. - T. 14. Central Asia. Vol. 1. Syr-Darya River Basin // Edited by I.A. Ilyin. - L.: Gidrometeeoizdat, 1969. - 439 c.

2. Engineering Hydrology. I.I. Levy, Gosenergoizdat, 1986.

3. General Hydrology, Hydrometeoizdat, 1982.

4. Ice appearance and the beginning of ice formation on rivers, lakes and reservoirs. Hydrometeoizdat, 1985.

5. Water economy and water management calculations. Stroyizdat, Bakhtiarov V.A., 1988.

6. Fundamentals of hydrometry. N.N. Pashkov, F.M. Dolgachev, M. Energoizdat, 1989.

7. Naryn River Basin (physical and geographical characterization) / Edited by R.D.Zabirov and V.A.Blagoobrazov. - Frunze: Izd. of the Academy of Sciences of the Kirg. SSR, 1960. - 229 c.

8. Goroshkov I.F. Hydrological calculations. - L.: Gidrometeeoizdat, 1979.

9. Zyryanov A.G. Dynamics of siltation of the Uchkurgan HPP reservoir and experience of sediment control // Hydrotechnical Construction, No. 1. - M., 1973, - P. 32-37.

10. Safety Assessment Report for the Toktogul and Uchkurgan Dams. Component C "Dam Safety and Reservoir Management". GEF Project. - Bishkek, 2001.

11. Construction norms and rules. SNiP 2.01.14-83. Determination of design hydrological characteristics. - Moscow: Stroyizdat, 1985.

12. Toktogul HPP on the Naryn River. Technical Design of Main Structures, Volume 1. Natural conditions. Book 2. Engineering and geological conditions. - 1036-T3. - SAO "Hydroproject", 1969.

13. TER of Uchkurgan HPP reservoir clearance on the Naryn River. Project No. 3805-17 / VO "SOYUZGIDROENERGOSTROY". - M., 1990.

14. Chu River Basin (physical and geographical characterization) / Edited by R.D.Zabirov and V.A.Blagoobrazov. - Frunze: Izd. of the Academy of Sciences of the Kirg. SSR, 1960. - 229 c.

15. Goroshkov I.F. Hydrological calculations. - L.: Gidrometeeoizdat, 1979.

16. Surface water resources of the USSR. - T. 14. Central Asia. Vol. 1. Syr-Darya River Basin // Edited by I.A. Ilyin. - L.: Gidrometeeoizdat, 1969. - 439 c.

17. Construction norms and rules. SNiP 2.01.14-83. Determination of design hydrological characteristics. - Moscow: Stroyizdat, 1985.

18. GTS. Edited by M.M.Grishin, M. Vysh.shk., 1989 Part 1 and 2.

19. Hydrotechnical constructions. - Rozanov N.P., M.: Stroyizdat.

20. Operation of hydromeliorative systems. Natalchuk M.F., Vysh. shk. 1989.

21. Engineering Geodesy. A.G.Grigorenko, M.I.Kiselev. M. Vysh.shk., 1989.

22. Laboratory practice on geodesy. M.I. Kiselev, M.Stroyizdat, 1990.

23. Hydrotechnical structures of complex structures. Edited by P.S. Neporozhnykh.

24. Hydropower engineering and integrated use of water resources of the USSR. Edited by P.S. Neporozhnykh M.: Energoizdat, 1982.

25. Regulation of river runoff. Pleshkov Y.F. L., Gidrometeoizdat, 1972.

26. Reservoirs of hydroelectric power plants. Avakyan A.B., Sharapov V.A. M.: Energia, 1986.

27. Reservoirs of the World. Avakyan A.B. M.: Energia, 1986.

28. Channels of Hydroelectric Power Plants. Korolev A.A. M., Gosenergoizdat, 1976.

29. Integrated use and protection of water resources. Zarubaev I.V., L. 1986.

30. Underground contour of hydraulic structures. Chugaev R.R. L., Energia, 1984.

31. Filtration calculations of hydraulic structures. Aravin V.I., Numerov S.N., M., Gosstroyizdat, 1990.

32. Hydraulics. Chugaev R.R. L., Energia, 1984.

33. Hydrotechnical structures. Chugaev R.R. L., Energia, 1984.

34. "Investigation of sediment deposition process in the Orto-Tokoi Reservoir" Mukanov T.A., No. 1-1, ISSN 2413-0869, Agency for Advanced Scientific Research (AASR), X International Scientific and Practical Conference: Modern Trends in the Development of Science and Technology, 2016, Belgorod, Russia, pp. 87-98.

appendix

Annex 1

Progress of siltation of Kambarata HPP-2 reservoir (average marks and water levels are shown at the end of the year)

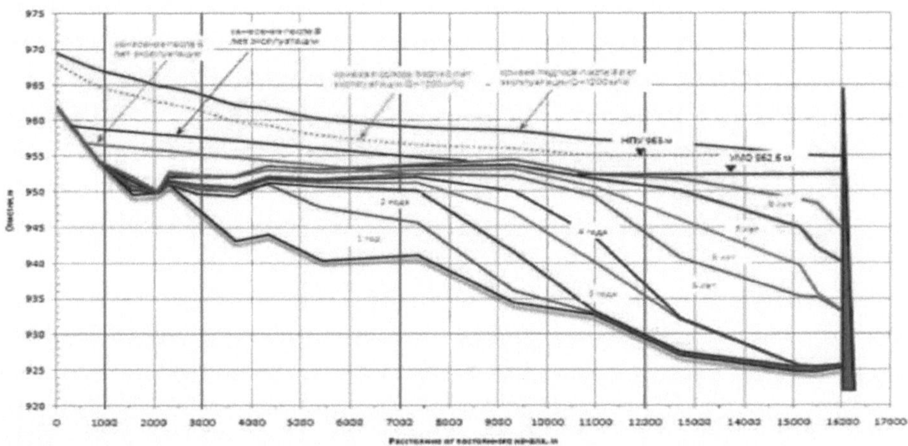

Annex 2

Siltation progress in the reservoir of Kambarata HPP-2 (dammed section of the reservoir)

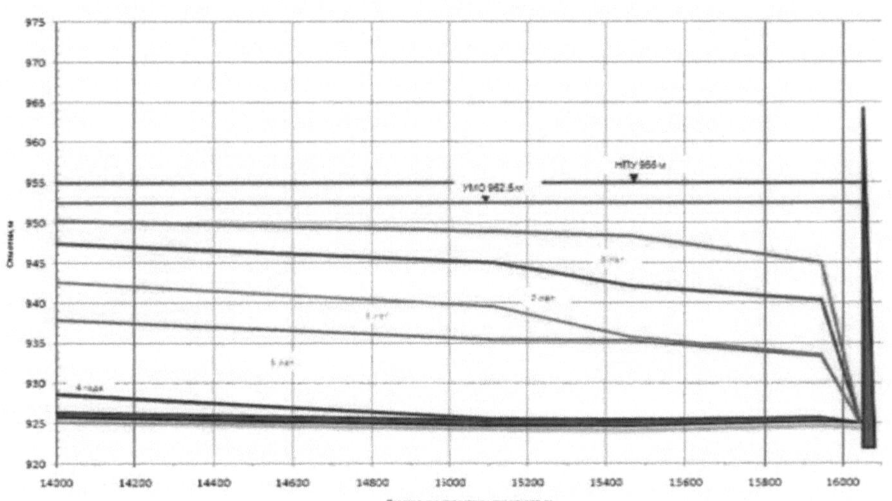

Annex

Volume curves of Kambarata HPP-2 reservoir volume

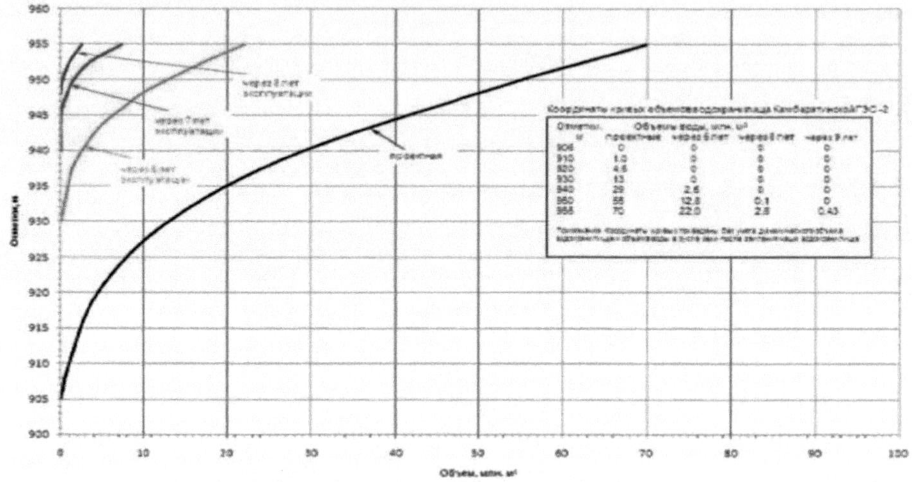

Annex

Progress of siltation of the Kambarata HPP-2 reservoir during drawdown up to 950.00 meters. (average bottom elevations and water levels are shown for the end of the year)

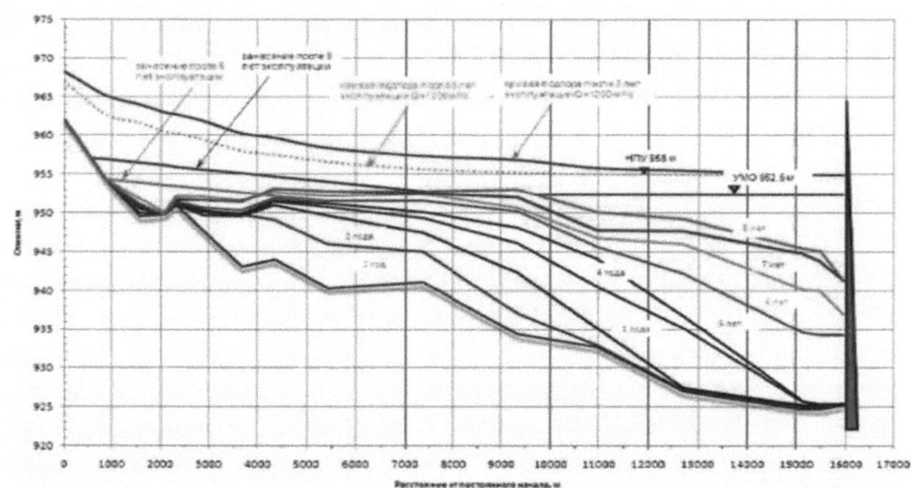

Annex 5

Siltation progress of Kambarata HPP-2 reservoir during drawdown up to 947.00 m. (average bottom elevations and water levels are shown for the end of the year)

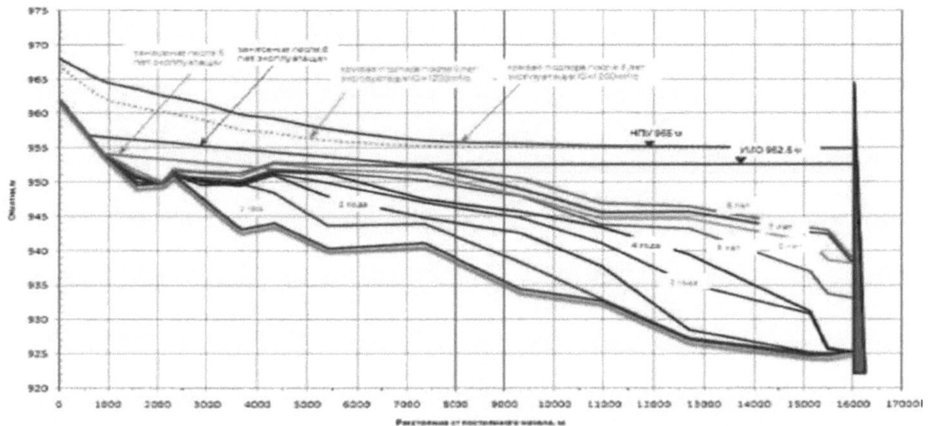

Annex 6

Cross section of the Naryn River at the entrance to the inlet channel of Kambarata HPP-2

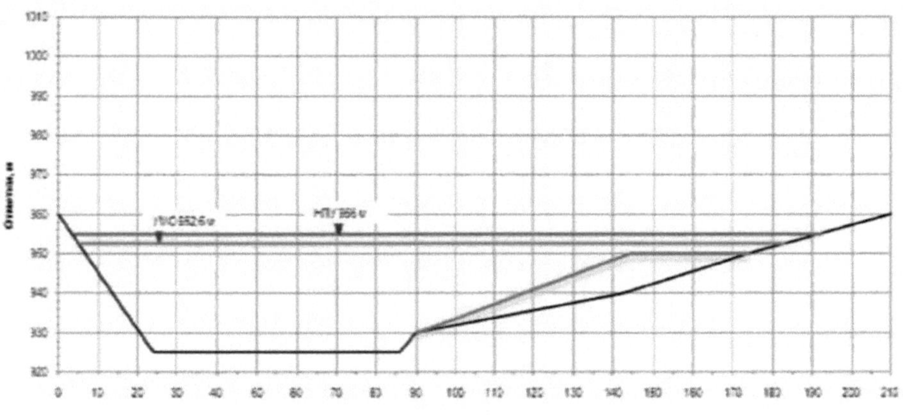

MIX
Papier aus verantwortungsvollen Quellen
Paper from responsible sources
FSC® C105338

Printed by Books on Demand GmbH, Norderstedt / Germany